STATISTICS FOR
SIX SIGMA MADE EASY!

Second Edition

WARREN BRUSSEE

McGraw Hill

New York Chicago San Francisco Lisbon
London Madrid Mexico City Milan New Delhi
San Juan Seoul Singapore Sydney Toronto

This publication is designed to provide accurate and authoritative information in regard to the subject matter covered. It is sold with the understanding that the publisher is not engaged in rendering legal, accounting, securities trading, or other professional services. If legal advice or other expert assistance is required, the services of a competent professional person should be sought.
> —From a Declaration of Principles Jointly Adopted by a Committee of the American
> Bar Association and a Committee of Publishers and Associations

This book is dedicated to my wife, Lois, who over the years has tolerated many of my idiosyncrasies, which include writing this book. Too bad it isn't a novel so she would actually read it!

CONTENTS

ACKNOWLEDGMENTS

Special thanks to Bonnie Burnick, a Six Sigma black belt, who encouraged me to write this book and who has provided invaluable assistance throughout the process.

Others who have given valuable input are Donald Brussee, Sharon Curtright, Jean-Patrick Ducroux, Roy McDonald, Cheri Sims, Russ Sims, and Christopher Welker.

Of course, none of these people are responsible for any errors or omissions, since the final decision on content was mine. Comments are encouraged by e-mailing Warren Brussee at wbsixsigma@aol.com.

INTRODUCTION

THE HISTORY OF SIX SIGMA

Motorola developed much of the Six Sigma methodology in the 1980s. The company was putting large numbers of transistors into its electronic devices, and every transistor had to work or the device failed. Therefore, Motorola decided that it needed a tighter quality criterion based on defects per million rather than the traditional defects-per-thousand measure. The initial quality goal for the Six Sigma methodology was no more than three defects per million parts.

Then, in the 1990s, Jack Welch, CEO of GE, popularized Six Sigma by dictating its use across the whole of GE. The resulting profit and quality gains that GE touted led to Six Sigma's being implemented in many large corporations. The claimed Six Sigma–generated savings are $16 billion at Motorola, $800 million at AlliedSignal, and $12 billion at GE in its first five years of use.

WHO SHOULD USE THIS BOOK

Manufacturing managers, engineers, and technicians. Implementing Six Sigma gives manufacturing and engineering teams a common language and a common approach to problem solving. No matter what skills people currently possess, the use of Six Sigma makes those skills more effective. *Statistics for Six Sigma Made Easy!* emphasizes using data to drive actions and get measurable results.

Sourcing. Six Sigma is equally valuable when it is applied to suppliers. The joint use of these tools makes your vendors an extension of your company. Suppliers are eager to participate, since they know that the Six Sigma process will both improve their product and strengthen the bond with their customer. These are almost always win-win situations.

Design engineers. Production problems are best solved in the design stage. Six Sigma uses data and customer input to assist in designing products and production equipment that are more likely to be problem-free. Of special interest are the chapter on tolerances, which emphasizes reality-based tolerances, and the various customer input tools.

Marketing and sales. Being able to demonstrate how your company uses Six Sigma tools to improve and control processes is a powerful marketing and sales tool. Many leading companies use Six Sigma, and a degree of prestige and perceived technical prowess comes with incorporating it. In addition, the tools can assist in spotting significant changes in demand or sales.

Accounting, software development, insurance, and other such applications. Although most of the initial Six Sigma applications have been in manufacturing, there is a growing awareness that these techniques work equally well in reducing costs or errors in other fields. They can be used to compare people, processes, companies, events, and other such data to spot significant differences or trends. Use the various customer input tools to benefit from the knowledge of everyone affected and to get maximum buy-in.

WHY *STATISTICS FOR SIX SIGMA MADE EASY!*

In a slow economy, companies no longer have the luxury of business as usual. Organizations must try to prevent quality issues, identify and solve problems quickly if they do occur, and drive down costs by increasing efficiencies. Six Sigma has proven its value in all these areas, and it is already in use at many large companies.

However, businesses of all sizes can benefit from the use of Six Sigma. Accounting firms, service companies, stockbrokers, grocery stores, charities, government agencies, suppliers of healthcare, and virtually any organization that is making a product or supplying a service can gain from the use of Six Sigma. In the healthcare field, for example, there is much controversy as to how to control rising medical costs. Yet there are few hard statistics related to the success of competing medical treatments. The need for good data analyses in the healthcare field makes Six Sigma a natural fit. As one specific example, in the United States, $8 billion per year is spent on various prostate cancer treatments. The costs of the various treatment options range from $2,000 to well over $50,000. Yet, according to a study by the Agency for Healthcare Research and Quality (February 4, 2008, U.S. Department of Health and Human Services, Agency for Healthcare Research and Quality), no one treatment has been proven superior to the others. The agency also noted that there is a lack of good comparative studies. Dr. Jim Yong Kim, a physician and the president of Dartmouth College, said that work done at Dartmouth showed great variation in outcomes among various hospitals and doctors, and he suggested that hospitals needed to learn from industry and start using processes like Six Sigma to improve and reduce the costs of healthcare delivery.

There are many books on Six Sigma. Most of them are just general overviews, with little detail on how to actually *use* the Six Sigma tools. *Statistics for Six Sigma Made Easy!* gently guides the user through the required statistics and enables someone to quickly apply Six Sigma tools to real-world problems.

Six Sigma is a structured methodology for solving problems with tools that can be applied to the problem-solving process. Its use generates insights that might not otherwise be obvious. Although it was initially designed to improve quality, many companies are now using Six Sigma to make cost-saving improvements. Implementers of this methodology are often called "green belts." This book simplifies the learning of Six Sigma and its application to green belt–level projects.

One company's green belt training in Six Sigma includes seven books, four software packages, and three weeks of classwork. It is very intensive in high-level statistics. Although this kind of course is excellent, not all companies or individuals want to commit to that level of instruction. *Statistics for Six Sigma Made Easy!* includes only the tools used by most successful Six Sigma practitioners. The only software package needed is Excel, and a brief review of using Excel to analyze data is included. The 13 formulas and 5 tables provided here will enable you to do all the Six Sigma work described in this book.

All required statistics are completely and simply detailed. Using these tools will enable a person to do much the same work as a green belt who has completed more extensive training, such as that described previously.

It is not necessary to master all the tools to become effective at utilizing Six Sigma. The application of even a few of them can have a strong impact on driving savings.

Many of the Six Sigma tools that are covered in this text are labeled as "simplified." This simplification in no way reduces their effectiveness. It just puts a degree of reality into the tools. In all cases, I give reasons for the simplifications and give reference texts for those who wish to use the traditional, nonsimplified tools.

This book can be used as a stand-alone or as a supplement to other Six Sigma texts.

TEACHING SIX SIGMA

Many of you who use this text will get involved in teaching Six Sigma. The people you will teach will have various educational backgrounds and various interest levels. I will share with you how I got involved in using and teaching Six Sigma, and finally writing this text. Perhaps my experiences will be valuable to you as you begin to teach this methodology.

I had already been at GE for many years and was managing a very successful engineering team when Six Sigma was introduced. GE initially had a limited number of very bright people trained by outside consultants who were deemed to be experts in Six Sigma. This group of newly trained people then put together a set of modules to be used to train the next group of people, which consisted mostly of managers.

Since these original trainers had no experience in actually using Six Sigma and had various degrees of ability in statistics, both the modules and the training were rather haphazard. The training consisted of two one-week sessions, which included an introduction to several software packages specific to Six Sigma.

In addition to holding these class sessions, GE brought in some outside consultants who covered additional Six Sigma tools and the corresponding software. Because of the software requirements, most people had to order new computers.

Therefore, there was a dichotomy in the training. The homegrown training modules, written and taught by people who were not experts in Six Sigma and had no application experience, were often weak. The training done by the consultants was often overwhelming. It was problematic that there was no practical text for teaching the use of the Six Sigma tools.

Over a period of months, this training was given to most of the engineers and to various other groups, such as marketing and sales. After taking these classes, the people were to start using Six Sigma; the goal was that within one year, everyone would complete two meaningful Six Sigma projects, document the savings, put in the necessary controls, and do a formal presentation. Those completing this would become "green belts."

Everyone was also to take a test at the end of one year on his or her competence in Six Sigma.

After most of the people on my engineering team had completed the classes, I asked them for feedback. At first I got the "they were OK" type of response. As I queried further, however, I found that the engineers had not truly learned or understood enough. The major weakness in the instruction was an assumption that the participants understood probability and statistics completely; little effort had been made to walk them through practical applications.

Since I took great pride in my team being one of the best, I decided to start teaching my team members a more practical version of Six Sigma. I proceeded to schedule four-hour Six Sigma sessions with my team every other week, in which I would cover some specific area of Six Sigma. The first sessions were a general review of statistics, emphasizing only what was needed to actually do Six Sigma work. After several months, the general manager asked me to start the training over with another team, which I did.

Everyone who attended the classes became a green belt by the one-year target, and the teams taking these classes beat all other teams on the test on Six Sigma that was given at the end of the year.

In the classes, I had to overcome the problem of a great diversity of skills and abilities. There were participants with two-year technical degrees, nontechnical degrees, engineering degrees, and even several doctorates in physics. I explained to all of them that I was going to start with the basics and move slowly, with applied problems as examples. I asked those who already had a good understanding of statistics to assist those who didn't. I made no attempt to identify the people who were in each group. This approach seemed to work. Even those who felt that they already had an understanding were surprised at how much they learned from these classes.

How did I prepare for the classes? The weekend before, I would get every resource I could find on the subject I was going to cover, then do my best to glean the important points and attempt to present them in an understandable way. Since the people I was teaching knew me well, feedback (positive and negative) was not an issue. This enabled me to fine-tune the course material. Hopefully this text will spare you the need for this level of preparation.

After several months of this, you could sense the pride building in these teams as their comprehension grew. They were even bragging to other teams that they were going to blow them away on the year-end test (which they did).

Another issue I had to address in these classes concerned a few very skilled individuals who over the years had done well without using Six Sigma. It was a difficult sell to get them to put much effort into learning and applying this methodology. However, these people were bright enough that they were able to acquire some degree of competence just by attending the classes.

Another facet of teaching Six Sigma comes into play when you start to actually use the methodology. You have to provide some training to all the people that you will be asking to contribute input or to help gather data. These people must get some feel for what the Six Sigma process does. An initial meeting of two to three hours is needed before involving them in the process. In the meeting, emphasize that Six Sigma needs input from knowledgeable people (them) and that the data that they will help gather will drive the decisions. Explain that you will be doing some statistical tests on the data; they don't need to understand all the details, but you will share the results. This means that you also need to have one or two additional meetings with them to keep them up to date.

There will be a few people within any group who will want to understand the Six Sigma process in far more detail. These people are extremely important to you, so you should have additional meetings for those who wish to understand more. They will be your best ambassadors!

There will be some fear that Six Sigma will be used to discipline employees in their jobs, or even to reduce the number of employees. Be very hesitant to do a Six Sigma project that has discipline or reduction of employees as a primary goal, since you will probably never get people to work with you again on Six Sigma. Obviously no one can guarantee that a reduction may not occur because of business conditions, but the project should not have that as a direct goal.

After several years of using Six Sigma and having a team that generated several millions of additional dollars of savings because of this methodology, I realized that I should go back and revisit the training method and material. This triggered the eventual writing of *Statistics for Six Sigma Made Easy!* Feedback from users of the first edition triggered the writing of this second edition.

LEAN SIX SIGMA

Traditional Six Sigma prioritizes reducing process variation, whereas Lean Six Sigma is mostly concerned with eliminating waste and streamlining production steps. However, there is a lot of overlap between traditional and Lean Six Sigma projects. With a traditional Six Sigma project, knowledge of the process flow is certainly needed to pinpoint areas of opportunity for reducing process variation. In fact, the process flow diagram is a primary traditional Six Sigma tool. When using this tool, it often becomes obvious that process flow can be streamlined and that this improvement can minimize both process variation and waste! For example, if raw materials are stored for excessive time periods, there can be subtle changes in those raw materials caused by water absorption, which can lead to both process variation and waste.

Similarly, when working on a Lean Six Sigma project, even though the emphasis is on streamlining process flow, excessive process variation caused by variable process flow will often be discovered; this will then give rise to projects involving both traditional and Lean Six Sigma. Both traditional and Lean Six Sigma use process flow diagrams as aids, but in Lean Six Sigma these process flow diagrams are often more detailed. They are then called "spaghetti maps" or "value stream maps." But they are still process flow diagrams! These will be discussed in more detail in the chapter on process flow diagrams.

Spaghetti maps and value stream maps help to identify areas of opportunity for Lean practitioners, but the means of improvement can vary greatly. In some cases, attempting to eliminate steps that don't add value is the best approach. Color coding of parts can reduce errors in assembly, but error proofing through parts redesign can often eliminate an assembly error completely. Parts placement, job redefinition, or combining of parts can both streamline a production step and eliminate waste. The Lean practitioner uses many of the traditional Six Sigma tools, such as QFDs and FMEAs, to identify potential solutions.

Many Lean Six Sigma practitioners find traditional Six Sigma intimidating because of its use of statistics. One of this book's goals is to enable people to do traditional Six Sigma without the fear that statistics often brings. Therefore, this book opens up another avenue of opportunity for Lean Six Sigma practitioners and makes their jobs more valuable.

PART I

**Overview of
the Six Sigma Process
and the DMAIC Road Map**

CHAPTER 1

Six Sigma Methodology and Management's Role in Implementation

What you will learn in this chapter is the basic structure and purposes of the Six Sigma process. You will also see how management's support of this methodology will ease its implementation and improve its likelihood of success.

Six Sigma Methodology

The Six Sigma methodology uses a specific problem-solving approach and specialized Six Sigma tools to improve processes and products. This methodology is data-driven, with a goal of reducing the number of unacceptable products or events.

The technical goal of the Six Sigma methodology is to reduce process variation to such a degree that the amount of unacceptable product is no more than 3 defects per million parts.

In most companies, the real-world purpose of Six Sigma is to make a product that satisfies the customer and minimizes supplier losses to the point where it is not cost-effective to pursue tighter quality.

DEFINITION

I am going to get into some Six Sigma terminology, but before I do, I want to explain what the Six Sigma methodology is about.

AVERAGE AND VARIATION

First, no one knows how to make anything "perfect." If you order fifty 1.000"-diameter ball bearings and then measure the bearings once you get them, you will find that none of them is exactly 1.000" in diameter. Some of them may be extremely close to 1.000", but if you measure them carefully, with a very good calibrated measuring device, you will find that the bearings do not measure exactly 1.000".

The bearings will vary from the 1.000" target for two reasons. First, the *average* diameter of these 50 bearings will not be exactly 1.000". The amount by which the average deviates from the target 1.000" is due to the bearing manufacturing process being *off-center*. Second, there will be a spread of measurements around the average bearing diameter. This spread of dimensions may be extremely small, but it will be there. This is due to the bearing process *variation*.

If the discrepancy caused by the combination of the off-center bearing process and the bearing process variation is small relative to your needs, then you will be satisfied with the bearings. However, if the discrepancy caused by the combination of the off-center process and the variation is large relative to your needs, then you will not be happy. The Six Sigma methodology strives to make the total effect of an off-center process and process variation small compared with the need (tolerance). This is illustrated in Exhibit 1-1.

Exhibit 1-1. Off-center and variation

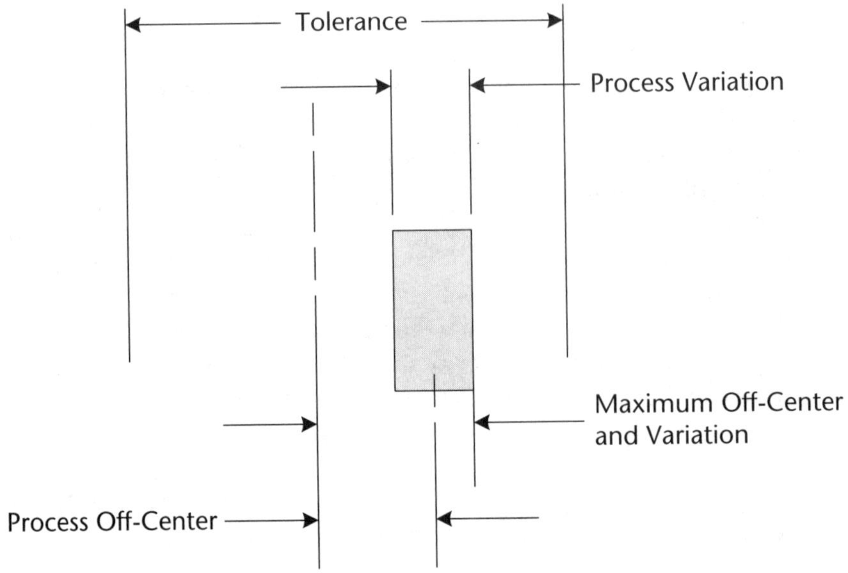

If you understand the general concepts that I just discussed, what follows is just terminology and detail.

SIGMA

One of the ways to measure the variation of a product or a process is to use a mathematical term called *sigma*. We will learn more about sigma and how to calculate this value as we proceed, but for now it is enough to know that the lower the value of sigma, the smaller the amount of process variation, and the higher the value of sigma, the greater the amount of process variation. Since the calculation of sigma is normally done on a computer or calculator, it is more important that you gain a sense that sigma is a measure of the data spread (variation) than it is to be too involved with the detailed actual calculation of sigma.

Ideally, the sigma value is low in comparison with the allowable tolerance on a part or process. If it is, the process variation will be small compared with the part or product tolerance that a customer requires. When this is the case, the process is "tight" enough that, even if it is somewhat off-center, the process produces products that are well within the customer's needs and specifications.

Most companies have processes that have a relatively large variation compared with their customers' needs (a relatively high sigma value compared with the allowable tolerance). These companies run at an average ±3-sigma level (a 3-sigma process). This means that 6 sigma (±3 sigma) fit between the tolerance limits (see Exhibit 1-2). The more sigma that fit between the tolerance limits, the better.

Exhibit 1-2. 3-Sigma Process

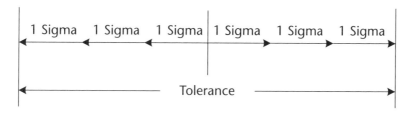

The sigma level is calculated by dividing the process's allowable tolerance (upper specification minus lower specification) by twice the process's sigma value, since the sigma level of a process is normally stated as a ± value.

DEFINITION

Process Sigma Level

$$\text{Process Sigma Level} = \pm \frac{\text{Process Tolerance}}{2 \times \text{Process Sigma Value}}$$

As an example, suppose a process machining shafts has the following measurements:

Sigma = 0.001"
Maximum allowable shaft diameter = 1.003"
Minimum allowable shaft diameter = 0.997"
So, the tolerance = 1.003" − 0.997" = 0.006"

We put these values into the given formula:

$$\text{Process Sigma Level} = \pm \frac{\text{Process Tolerance}}{2 \times \text{Process Sigma Value}}$$

$$\text{Process Sigma Level} = \pm \frac{0.006"}{2 \times 0.001"}$$

$$\text{Process Sigma Level} = \pm 3$$

So, this process is running at ±3 sigma, or, in the terms of Six Sigma, this is a 3-sigma process.

As you will see later in the book, a ±3-sigma process generates 99.73 percent good product, or 997,300 good parts out of every 1,000,000 parts produced. This means that out of every 1,000,000 parts produced, there are 2,700 defective parts (or 2,700 input errors per 1,000,000 computer entries, or some other measure). These defects are very costly, as they cause scrap, rework, returns, and loss of customers. Eliminating this lost product has the potential to be a very profitable "hidden factory," because all the costs and efforts have been put into the defective product, but it's unusable.

Some companies, like aircraft manufacturers, attempt to run very critical parts at a 7-sigma level. This means that the variation is so low that ±7 sigma fit between the specification's high and low limits. This targeted extremely low defect level (low process variation compared with the allowable tolerance) is sought because of the catastrophic potential of a defect (a part being outside tolerance).

A Six Sigma process runs with a variation such that ±6 sigma (including some process drift), or 12 sigma, fit within the tolerance limits. This will result in three defects per million parts produced. This was the original quality goal of this methodology, and it's how the name "Six Sigma" became associated with the methodology. This extremely low defect incidence is not required in most real-world situations, and the cost of getting to that quality level is usually not justified. However, getting the quality to the level where the customers are extremely happy and the suppliers' losses are very low is generally a cost-effective goal.

Companies that embrace the Six Sigma philosophy train people to various skill and responsibility levels and assign the following titles.

Green Belt

A Six Sigma green belt is the primary implementer of the Six Sigma methodology. He earns this title by taking classes in Six Sigma, demonstrating a competence on Six Sigma tests, and implementing projects using the Six Sigma tools.

Black Belt

A Six Sigma black belt has Six Sigma skills sufficient to allow her to act as an instructor, mentor, and expert to green belts. A black belt is also competent in additional Six Sigma tool-specific software programs and statistics.

Master Black Belt

A Six Sigma master black belt generally has management responsibility for the Six Sigma organization. This could include setting up training, measuring its effectiveness, coordinating efforts with the rest of the organization, and managing the Six Sigma people (when Six Sigma is set up as a separate organization).

DEFINITION

Much of Six Sigma is not new. The Six Sigma methodology includes elements from SPC (statistical process control), the scientific method of problem solving, and procedures for incorporating expert knowledge. Practical application of statistics and probability is inherent in the Six Sigma process. However, the Six Sigma methodology brings all these elements together in a synergistic and disciplined fashion that has proven to be effective in driving process improvement. The use of Six Sigma doesn't change someone's job; it just makes him more effective in doing that job.

IMPLEMENTING SIX SIGMA

There are several elements and possibilities involved in implementing Six Sigma.

Commitment of Top Management

The easiest way to implement Six Sigma in an organization is to have complete commitment from top management. This commitment should include companywide communications explaining the process and its goals, with some explanation of the reasons why the company is investing the time and energy into implementing the Six Sigma methodology. This buy-in demonstrates to the whole company that management believes in this methodology, meaning that the required investment in people and training will happen, and that the program will have everyone's active participation. When incorporating Six Sigma, many companies start with outside consultants or instructors and then make the transition to in-house people as trainers.

Six Sigma Separate

Some companies set up Six Sigma as a separate organization, which then services the rest of the company. As a separate organization, the Six Sigma people work in parallel with the various groups that are already in place, identifying and implementing Six Sigma projects in addition to whatever projects the groups have already defined.

   The advantage of this approach is that using it to implement Six Sigma means that fewer people need to be trained initially and that the effect of the methodology can be tracked more readily. The downside of this approach is that the separate Six Sigma organization is often looked at as a group of prima donnas, with its own set of agendas. This causes some resentment among others in the organization and stifles cooperation. It also discourages the input of experts into Six Sigma projects, since many of these experts feel threatened. There can also be some feeling that current ideas are being "stolen" and then labeled as Six Sigma.

Six Sigma Integrated

An alternative approach is to incorporate Six Sigma as part of the organization, not as a separate entity. Six Sigma then becomes an integral part of everyone's job, with a relatively few highly trained Six Sigma people being used as reference instructors. This is a somewhat more difficult way to implement Six Sigma, because of the large number of people that have to be trained, but a common Six Sigma language and philosophy will then permeate the organization. As the Six Sigma methodology unfolds in the coming chapters, it will be seen that Six Sigma is helpful to everyone in the organization, and therefore it should become an integral part of everyone's job.

Six Sigma from the Bottom Up

Sometimes some high-level managers and/or people throughout the company feel that the company can't afford the training, software, computer upgrades, and other elements needed to implement Six Sigma. Managers may also be dubious as to whether the skill levels that exist in the company will support the perceived high technical competence required. In these cases, complete management buy-in is unlikely. Although a complete commitment from management is the easiest way to implement Six Sigma, it is possible for individuals or teams to start this process from the bottom up.

Six Sigma Tools

Many of the Six Sigma tools will work independently. The tools can also be simplified and do not require the rather esoteric special software that is often associated with Six Sigma. Excel is sufficient for all calculations and graphs. Even the implementation of two or three Six Sigma tools can make a measurable difference in a company's performance! There is no need to try to implement the whole methodology at once.

Of course, an individual who is planning to use Six Sigma should first review her implementation plans with her manager, but generally a manager will not discourage this extra effort. Usually there will be a caveat from the manager not to incur any additional costs and not to delay any ongoing projects.

After someone has demonstrated the success of the Six Sigma method, others will often follow her lead. Then tool use and training can expand. Although this takes individual initiative on the part of the person introducing Six Sigma, it is a great way to get noticed and to truly influence a company's success. Even when a company is supportive of Six Sigma, it generally takes a few dynamic individuals to lead the effort.

WHAT WE HAVE LEARNED IN CHAPTER 1

1. The Six Sigma methodology uses a specific problem-solving approach and specific Six Sigma tools to improve processes and products.
2. People with expertise in Six Sigma are called green belts, black belts, or master black belts.

3. The name Six Sigma came from the goal of reducing defects to 3 parts per million, which is ±6 sigma (including some process drift).

4. Most companies produce at an average ±3-sigma quality level (99.73 percent good product), which generates 2,700 defects per million parts. For most real-world situations, this level of defects is excessively high, but the optimum quality level is usually not as tight as ±6 sigma. A realistic goal is to make a product that satisfies the customer and minimizes supplier losses to the point where it is not cost-effective to pursue tighter quality.

5. Excessive defects are very costly, leading to scrap, rework, returns, lost customers, and other such losses. This lost product has the potential to be a profitable "hidden factory."

6. The easiest way to implement Six Sigma in an organization is with complete commitment from top management.

7. Although complete corporate commitment is desirable, many of the Six Sigma tools can be used independently to make substantial improvements. This approach can lead to bottom-up acceptance of Six Sigma. *It is not necessary to use all the tools to have a measurable effect on reducing defects.*

8. Some companies set up Six Sigma as a separate organization. This can cause some animosity. Another approach is to incorporate Six Sigma into the current organization, as an integral part of everyone's current job.

9. There are task-specific software programs for many of the Six Sigma tools, but these are not required to begin implementing Six Sigma.

10. The Six Sigma tools can be simplified to make them more practical without significantly reducing their value.

RELATED READING

Dick Smith and Jerry Blakeslee, *Strategic Six Sigma: Best Practices from the Executive Suite* (New York: John Wiley, 2002).

Peter S. Pande, Robert P. Neuman, and Roland R. Cavanagh, *The Six Sigma Way: How GE, Motorola, and Other Top Companies Are Honing Their Performance* (New York: McGraw-Hill, 2000).

Mikel J. Harry and Richard Schroeder, *Six Sigma: The Breakthrough Management Strategy Revolutionizing the World's Top Corporations* (New York: Random House/Doubleday/Currency, 1999).

CHAPTER 2

DMAIC: The Basic Six Sigma Road Map

What you will learn in this chapter is the DMAIC problem-solving approach that green belts use. This is the road map that is followed for all projects and process improvements. Which tools are used and what statistics are needed are dictated by the particular project. Appendix A gives a Six Sigma Statistical Tool Finder Matrix to assist in picking the correct tool.

DMAIC Problem-Solving Method

DMAIC (Define, Measure, Analyze, Improve, Control) is the Six Sigma problem-solving approach that green belts use. This is the road map that is followed for all projects and process improvements, with the Six Sigma tools being applied as needed.

D: Define. This is the overall definition of the problem. It should be as specific as possible.

M: Measure. Accurate and sufficient measurements and data are needed.

A: Analyze. The measurements and data must be analyzed to ensure that they are consistent with the problem definition and to identify a root cause. A problem solution is then identified. Sometimes, based on the analysis, it is necessary to go back, restate the problem definition, and start the process over.

DEFINITION

I: Improve. Once a solution is identified, it must be implemented. After the solution has been implemented, the results must be verified with independent data.

C: Control. A verification of control must be implemented. A robust solution (like a part change) will be easier to keep in control than a qualitative solution.

As we learn to use each tool in the following chapters, I will refer back to its use in the DMAIC process.

D: Define

A problem is often initially identified very qualitatively:

- "The customer is complaining that the quality of the bearing races has deteriorated."
- "The new inventory tracking software program keeps crashing."
- "The losses on line 3 seem higher."

Before we can even think about possible solutions, the problem must be defined more specifically. Only then can meaningful measurements or data be collected. After some additional definition, the previous examples could appear like this:

- "The inside diameter of the MQ18 bearing race became more varied starting in week 14."
- "When the number of inventory items exceeds 1,000, the inventory-tracking software crashes several times per day."
- "The number of line 3 product being scrapped for loose wiring has doubled in the last week."

If there are quantitative values available, such as the specific measurements related to the bearing diameter, they should be included in the problem definition. The more specific the initial problem definition, the better.

Getting a good definition of the problem may be as simple as talking to the customer. This text includes several excellent tools for quantifying customer input. However, developing an improved definition will often require much more effort than that. Some preliminary measurements may have to be taken to be sure that there even *is* a problem. It may be necessary to verify measurements and calculate sample sizes to ensure that we have valid and sufficient data. Sometimes the resultant measurements and analysis will show that the initial problem definition was erroneous, and you then have to back up and formulate another definition.

M: Measure

Once the problem has been defined, we must decide what additional measurements must be taken to quantify it. We will discuss several tools that will help you identify the key process input variables to be considered and/or measured.

Samples must be sufficient in number, random, and representative of the process that we wish to measure.

A: Analyze

Now we have to see what the data are telling us. We have to plot the data to understand the character of the process. We must decide whether the problem as defined is "real," or whether it is just a random event without an assignable cause. These data will also be the base against which we will measure any implemented improvement. We may also have to measure appropriate key process input variables.

I: Improve

Once we understand the root cause of the problem and have quantitative data, we identify alternative solutions. Tests may be required to understand any interaction between or among the input variables. Tolerances have to be examined. We analyze the error contributed by each component to see if one component is causing most of the error.

We then implement the solution and verify the predicted results.

C: Control

Quality control data samples and measurement verification can be scheduled. Control charts can be implemented to help the operator keep the process in control. Updated tolerances should reflect any changes.

USING DMAIC

It is strongly recommended that you follow all the DMAIC steps when problem solving. In particular, don't make major process changes without going through all the DMAIC steps. Remember, trying to fix something without working through all the applicable steps may cause you to spend more time responding to the resultant problems than you would have needed if you had taken the time to do it right!

Not only is the DMAIC road map useful for problem troubleshooting, but it also works well as a checklist when you are doing a project. In addition to any program management tool that is used to run a project, it is often useful to make a list of the Six Sigma tools that are planned for each stage of the DMAIC process as the project progresses. This Six Sigma tool checkoff list should be reviewed and updated regularly as the project progresses.

Appendix B gives an example of a Six Sigma tool checkoff list that relates to the case study given here. By the time you get to Appendix B, you will be familiar with the specific Six Sigma tools that are shown in the checkoff list example.

CASE STUDY: USING THE DMAIC PROCESS FOR PROJECT CONTROL AND REVIEW

Six Sigma was implemented at a company, and the engineering team began to follow the DMAIC process religiously. Not only did the team's project performance improve, but also the customer plants now felt that they were a part of the process. This is because many of the Six Sigma tools required input from the customers. Even when there were temporary issues, everyone felt that he was part of the problem and therefore felt that he should be part of the solution.

Here's a comment from one of the program managers, who is currently managing 17 diverse programs in several plants, with a total project value of over $2,000,000:

I can't imagine ever going back to managing programs without the DMAIC Six Sigma process. Not only are more of my programs meeting their goals, but also they are easier to manage now that the plants, and my manager, can track the steps in each project. Every project is reviewed using the DMAIC format. At the start of every project, a list of Six Sigma tools that is specific to that program is identified. At every program review, this Six Sigma tool checkoff list is reviewed to make sure that every element is being followed. There are no surprises. The DMAIC process minimizes the "panic" and catch-up that often accompanied projects before Six Sigma.

The advantage of preparing the Six Sigma tool checkoff list at the start of the project is that it is less likely that a tool will not be used.

In the middle of a project, when all sorts of things are happening, including the pressure of other projects, it is easy to skip a tool. That is less likely to happen when you know that the tool is already on a review list against which you will be measured.

As an example, on a project that was completed a year ago, we were in the last steps of verifying that a new inspection gauge was working as designed. It had already passed the tests with sample products, and it was then to be tested on actual production parts, with its performance measured against historical data.

The project was running a few days late, and the production manager wanted to skip the test with production products, since he was sure the test with the sample products was sufficient. The project would then be on time. I insisted that we do the test as planned, even though that meant that the project would be perhaps a week late.

The inspection device failed the test with production products. The device was rejecting a statistically significant lower number of production products for rim diameters than historical (population) rejects using the old gauge. After a week of testing production product with the new and old devices, it was discovered that there was a type of rim diameter that had a small amount of distortion, and the new inspection gauge was not able to measure the diameter correctly because of this distortion. No one had been aware of this particular distortion because the old dial indicator gauge was not sensitive to it. After a very minor modification of the new gauge, however, rim diameters having this distortion were readable. Products with the distorted rims were added to the sample products, the tests were redone, and the new gauge then passed with flying colors.

The inspection device was two weeks late because of the time needed to identify and fix the problem, and to redo the tests. However, if the test with production products had *not* been run, the problem would not have been found until the customer started finding more products with rim diameter defects. Then we would have had a very angry customer, would have had a great deal of product to reinspect, and would have lost faith in the Six Sigma process! As it was, the two-week delay was forgiven when everyone realized how the Six Sigma tool checkoff list had saved us *all* a lot of grief!

WHAT WE HAVE LEARNED IN CHAPTER 2

1. The DMAIC process (Define, Measure, Analyze, Improve, and Control) is the process road map that Six Sigma green belts use to solve problems.
2. The Six Sigma tools are used in different steps in the DMAIC process. The particular project or problem dictates which tools are used in the DMAIC process and at what point.
3. Not only is the DMAIC process useful as a problem-solving guide, but it also can be used as a standardized format for project reviews. A Six Sigma tool checkoff list is an effective way to make sure that the applicable tools are identified and used. An example of a Six Sigma tool checkoff list is shown in Appendix B.
4. The use of Six Sigma—both the DMAIC process and the Six Sigma tools—gets involvement and buy-in from both customers and management.
5. There is less panic and more control when this methodology is followed. Effectiveness improves measurably, as do job satisfaction and reward.

RELATED READING

Rath and Strong Management Consultants, *Rath & Strong's Six Sigma Pocket Guide* (Lexington, MA: Rath & Strong/Aon Consulting Worldwide, 2000).

Thomas Pyzdek, *The Six Sigma Handbook, Revised and Expanded: A Complete Guide for Green Belts, Black Belts, and Managers at All Levels* (New York: McGraw-Hill, 2003).

PART II

**Qualitative
Six Sigma Tools**

CHAPTER 3

Simplified QFD

What you will learn in this chapter is that what a customer really needs is often not truly understood during the design or change of a product, process, or service. A simplified QFD, if done carefully, will minimize issues arising from this lack of understanding.

QFD originally stood for *quality function deployment*. Years ago, when quality departments were generally much larger than they are now, quality engineers were "deployed" to the customers to rigorously probe a customer's needs and then create a series of forms that made the transition from those customer needs to a set of actions that the supplier could take. The simplified QFD attempts to accomplish the same task in a condensed manner.

What is presented here is a simplified version of the QFDs that are likely to be described in many Six Sigma classes. Some descriptions of these traditional QFDs and the rationale for the simplification will be given later in this chapter. The simplified QFD is usually used in the Define or the Improve step of the DMAIC process.

A simplified QFD is a Six Sigma tool that does not require any statistics. However, it is usually necessary to do a simplified QFD to understand what actions are needed to address a problem or implement a project. The specific actions that are identified in the QFD, or in any of the other qualitative tools, are often what trigger the application of the statistically based Six Sigma tools.

Simplified QFD

The simplified QFD converts customer needs into prioritized actions.

  Here are some examples of how a QFD is used.

Manufacturing. Use the simplified QFD to get input from customers on their needs at the start of every new design or before any change in process or equipment.

Sales and marketing. Before any new sales initiative, do a simplified QFD, inviting potential customers, salespeople, advertisement suppliers, and other such groups to give input.

Accounting and software development. Before implementing a new program language or software package, do a simplified QFD. A customer's input is essential for a seamless introduction.

Receivables. Do a simplified QFD on whether your approach to collecting receivables is optimized. In addition to those who are directly involved in collections, invite customers who are overdue on receivables to participate. (You may have to give them some debt relief to get their cooperation.)

Insurance. Do a simplified QFD with customers to see what they look for when they pick an insurance company or what it would take to make them switch.

Many product, process, and service issues are caused by not incorporating inputs from customers and/or from suppliers of components and raw materials early in a design or process change. Often the manufacturers' decision makers just assume that they and the people from whom they source the components and raw materials already know what the customers want.

The customers in this case include everyone who will touch the product while or after it is made. This includes employees in production, packaging, shipping, and sales as well as the end users. They are all influenced by any design or process change. The people who operate equipment, who do service work, or who implement can be both customers and suppliers.

The most difficult (and important) step in doing any QFD is getting the suppliers, operators, and customers together to prepare the required QFD form(s). Every group that is affected by the project should be represented. The desires of one group will sometimes lead to limitations on others, and simultaneous discussions among the factions will often identify options that were not previously considered, to arrive at the best possible overall solution. As you read the following details, refer to the simplified QFD form (Exhibit 3-1) to see the application.

Exhibit 3-1. Simplified QFD example and form

License Plate Holder Option for Luxury Automobiles
Ratings:
5 = Highest
1 = Lowest (or negative number)

Customer Needs	Ratings	Metal Cast Rim	Plastic Cast Complete	Stamped Steel Rim	All Holes Already In	Optional Punched Holes	Gold/Silver Plating	Hex/Slotted Plastic Screws	Hex/Slotted Plated Steel Screws	Plastic Lens, Separate	Tempered Glass Lens, Separate
Embossed Name	5	5 / 25	5 / 25	3 / 15	0 / 0	0 / 0	0 / 0	0 / 0	0 / 0	0 / 0	0 / 0
Place for Dealer Name	5	5 / 25	5 / 25	5 / 25	0 / 0	0 / 0	0 / 0	0 / 0	0 / 0	0 / 0	0 / 0
Must Hold All State Plates	5	5 / 25	5 / 25	5 / 25	5 / 25	5 / 25	0 / 0	0 / 0	0 / 0	0 / 0	0 / 0
Solid Feel	3	5 / 15	3 / 9	2 / 6	0 / 0	0 / 0	0 / 0	1 / 3	4 / 12	2 / 6	5 / 15
Gold/Silver Option	2	5 / 10	5 / 10	3 / 6	0 / 0	0 / 0	5 / 10	5 / 10	5 / 10	0 / 0	0 / 0
Easy to Install	4	3 / 12	5 / 20	3 / 12	5 / 20	3 / 12	0 / 0	5 / 20	3 / 12	3 / 12	2 / 8
Corrosion Resistance	4	3 / 12	5 / 20	2 / 8	0 / 0	0 / 0	3 / 12	5 / 20	3 / 12	4 / 16	5 / 20
Light Weight (for MPG)	1	1 / 1	4 / 4	5 / 5	0 / 0	0 / 0	0 / 0	5 / 5	2 / 2	5 / 5	1 / 1
Luxury Look	4	5 / 20	2 / 8	1 / 4	2 / 8	4 / 16	5 / 20	1 / 4	5 / 20	2 / 8	5 / 20
No Sharp Corners	5	4 / 20	5 / 25	2 / 10	4 / 20	2 / 10	0 / 0	5 / 25	3 / 15	5 / 25	2 / 10
Transparent Lens	4	0 / 0	3 / 12	0 / 0	0 / 0	0 / 0	0 / 0	0 / 0	0 / 0	3 / 12	5 / 20
Last 10 Years	5	4 / 20	4 / 20	3 / 15	0 / 0	0 / 0	3 / 15	5 / 25	4 / 20	3 / 15	5 / 25
Low Cost	2	1 / 2	5 / 10	4 / 8	5 / 10	2 / 4	1 / 2	5 / 10	3 / 6	4 / 8	2 / 4
Keep Appearance 10 Years	4	4 / 16	3 / 12	2 / 8	0 / 0	0 / 0	2 / 8	5 / 20	3 / 12	3 / 12	5 / 20
Groupings Totals Priorities		203 1	225	147	83 3	67	67 4	142 2	121	119 NA	143 NA

SIMPLIFIED QFD INSTRUCTIONS

The simplified QFD form is a way of quantifying design options, always measuring these options against customer needs. The first step in preparing the simplified QFD form is to make a list of the customer needs. Then, assign a value of from 1 to 5 to each need:

- 5 means that it is a critical or a safety need, a need that must be satisfied.
- 4 means that it is a very important need.
- 3 means that it is highly desirable.
- 2 means that it is a "nice to have."
- 1 means that it is a "wanted if it's easy to do."

You can use a more elaborate rating system, but if you do, you will find that you are spending too much time assigning numbers! The customer needs and ratings are listed down the left side of the simplified QFD form.

Across the top of the simplified QFD form are potential actions to address the customer needs. Note that the customer needs are often expressed qualitatively (easy to use, won't rust, long life, and the like), whereas the design action items listed will be more specific (tabulated input screen, stainless steel, sealed roller bearings, and the like). Under each design action item and opposite each customer need, you will determine a value (1 to 5) that indicates how strongly that design item addresses the particular customer need:

- 5 means that it addresses the customer need completely.
- 4 means that it addresses the customer need well.
- 3 means that it addresses the customer need somewhat.
- 2 means that it addresses the customer need a little.
- 1 means that it addresses the customer need very little.
- 0 or blank means that it does not affect the customer need.
- A negative number means that it is detrimental to that customer need. (A negative number is not that unusual, since a solution to one need sometimes interferes with another need!)

Put the rating in the upper half of the block beneath the design item and opposite the need. Then, multiply the design rating times the value assigned to the corresponding customer need. Enter this result in the lower half of the square under the design action item rating. These values will have a possible high of 25.

Once all the design items have been rated against every customer need, add up the values in the lower half of the boxes under each design item and enter the sums into the Totals row at the bottom of the sheet. The solutions with the highest values are usually the preferred design solutions to address the customer needs.

Upon reviewing these totals, someone may feel that something is awry and want to go back and look again at some ratings or design solutions. This second (or third) review is extremely valuable. Also, the customer 5 ratings should be discussed one at a time to make sure that they are being addressed sufficiently.

Exhibit 3-1 is an example of a simplified QFD. The simplified QFD form can be done by hand or in Excel. In either case, the building of the form and the rating of each item should be done "live" in the meeting to get maximum interaction and participation.

In the simplified QFD form, near the bottom, design action items are grouped when only one of the options can be done. The priorities within a group are only among the items in that group.

In this case, the priorities showed that the supplier should cast a plastic license plate cover with a built-in plastic lens. This precludes the need for a separate lens, which is why NA (not applicable) is shown for both items in that grouping. The unit should be mounted using plastic screws, with holes for all plates cast in. Gold or silver plating is an option that can be applied to the rim of the plastic.

Note that some of the luxury items (like steel casting) weren't picked because other factors were deemed to be more important. Having the customers there for these discussions is very critical, so that they don't feel that the supplier is giving them something other than what they really want. Often customers will start out with a wish list of items that is then trimmed down to the critical few.

The form can be tweaked to make it more applicable to the product, process, or service that it is addressing. What is important is getting involvement from as many affected parties as possible and using the simplified QFD form to drive the direction of the design.

The simplified QFD form should be completed for all new designs or for process modifications. The time and cost involved in holding the required meetings will be more than offset by making the correct decisions up front, rather than having to make multiple changes later.

CASE STUDY: A SIMPLIFIED QFD ON A TEST CUTTING MACHINE

A simplified QFD was being prepared to determine what adjustments were to be incorporated on a test in-line cutting machine for extruded tubing. The intent of this project was to instrument a cutting machine that would be sufficient to test all reasonable combinations of settings and identify an optimum cutting process. The cutting machines currently were leaving a rough end on the tubing as they cut it. This caused excessive material losses.

Future production machines would be designed with features dictated by the results found using the test cutting machine. The purpose of the simplified QFD was to identify which items were to be tested on this test in-line cutter. Then, equipment design engineers would have to design that test capability into the test machine.

In attendance at this simplified QFD meeting were several PhDs who had studied the process, equipment design and process engineers, customers who emphasized what issues had to be addressed, and some operators and maintenance people who were knowledgeable about the current cutting equipment.

The machine operators were quiet until the end of the meeting, when they became insistent that an additional support spring be added. None of the "experts" felt that this additional support spring was needed, and they opposed adding it because the operators could not logically explain the rationale for the spring. The operators' experience, however, made them adamant that the support spring was needed. Since the operators were so insistent and the worst thing that would happen was that the added spring would prove not to be needed, it was decided to include the additional support spring against the advice of the "experts."

When the test was run to find the optimum settings, it was found that this additional support spring was critical. The cutting process was very unstable without this support spring being optimized. Without the simplified QFD, this spring would not have been incorporated, and the process optimization would not have been as successful.

The knowledge gained from this piece of test equipment was incorporated into the design of 12 in-line extruded tubing cutters, saving $1,500,000 per year.

TRADITIONAL QFDs

A traditional QFD, as taught in most classes on Six Sigma, is likely to be one of the following.

The first is a QFD consisting of four forms. The first form is called the House of Quality. This form covers product planning and competitor benchmarking. The second form is Part Deployment, which shows key part characteristics. The third form shows Critical-to-Customer Process Operations. The fourth covers Production Planning.

Two other types of QFDs are the Matrix of Matrices QFD, consisting of 30 matrices, and the Designer's Dozen, consisting of 12 QFD matrices.

Needless to say, these other QFDs take much more time and effort than the simplified QFD. Meetings to complete the traditional QFDs generally take at least four times as long as the meetings required for the simplified QFD on an equivalent project.

Are the traditional QFDs worth the extra effort over that required by the simplified QFD? On very large and very complex programs, perhaps yes. However, the benefit of being able to use the simplified QFD on *every* project or change, which is not realistic with the more complex QFDs, gives it a decided advantage. The customers' inputs are needed on *all* levels of projects!

The major benefit of any QFD comes from getting the input of everyone affected. If the form is too complex, or if the meeting required to complete the form is too long, people lose focus, their eyes begin to blur, and the quality of the input diminishes. The simplified QFD is designed to get the needed input with the minimum of hassle! Simplified QFD meetings are often only two or three hours long.

WHAT WE HAVE LEARNED IN CHAPTER 3

1. The simplified QFD is usually used in the Define or Improve step of the DMAIC process.
2. Many product, process, and service issues are caused by not incorporating inputs from customers and/or from suppliers of components and raw materials early in a design. Often the manufacturer just assumes that what the customers really want is already known.
3. The use of a simplified QFD can minimize a lot of issues before a design or modification is implemented.

4. Everyone who is affected by the project—such as operators, suppliers, users, engineers, maintenance staff, and customers—must participate in generating the simplified QFD form.
5. The simplified QFD should be prepared as early in a project as possible.
6. Revisit the simplified QFD's priority results several times to make sure that they truly reflect the group's intent.
7. The cost of doing a simplified QFD will be more than offset by the benefits of a superior design with fewer modifications required.
8. The simplified QFD should be used on *every* new product and *every* product or process change.
9. There are more complex and detailed QFDs that may be worth considering for very large and very complex programs. However, the effort and people required for these QFDs usually preclude their being used on smaller designs and changes. The simplified QFD is extremely practical, since it *can* be used be used on *all* designs and changes.

RELATED READING

Shigeru Mizuno and Yoju Akao, eds., *QFD: The Customer-Driven Approach to Quality Planning and Deployment* (Tokyo: Asian Productivity Organization, 1994).

Lou Cohen, *Quality Function Deployment: How to Make QFD Work for You* (Upper Saddle River, NJ: Prentice Hall PTR, 1995).

Mark J. Kiemele, Stephen R. Schmidt, and Ronald J. Berdine, *Basic Statistics: Tools for Continuous Improvement*, 4th ed. (Colorado Springs, CO: Air Academy Press, 1997).

CHAPTER 4

Simplified FMEA

What we will learn in this chapter is that on any project, there can be collateral damage to areas outside the project. A simplified failure modes and effects analysis (FMEA) will reduce the likelihood of such damage. A simplified FMEA will generate savings largely through cost avoidance, and it is usually used in the Define or Improve step of the DMAIC process.

As was true for simplified QFDs, as presented in the prior chapter, the simplified FMEA will be less complex than the FMEA taught in most Six Sigma classes. A brief discussion of the traditional FMEA and the reasons for the simplification comes later in the chapter.

Note that the format for the simplified FMEA is very similar to that used for the simplified QFD. This is intentional, since the goal is to use *both* analyses on *every* new product or change. Since many of the same people will be involved in both the QFD and the FMEA, the commonality of the two forms simplifies the task.

Simplified FMEA

Manufacturing. Before implementing any new design, process, or change, do a simplified FMEA. An FMEA converts qualitative concerns into specific actions. You need input on what possible negative effects could occur.

Sales and marketing. A change in a sales or marketing strategy can affect other products or lead to an aggressive response by a competitor.

APPLICATIONS

A simplified FMEA is one way to make sure that all the possible ramifications are understood and addressed.

Accounting and software development. The introduction of a new software package or a different accounting procedure sometimes causes unexpected problems for those affected. A simplified FMEA will reduce unforeseen problems and trauma.

Receivables. How receivables are handled can affect future sales. A simplified FMEA will help everyone involved to understand the concerns of both customers and internal salespeople and to identify approaches that minimize future sales risks while reducing overdue receivables.

Insurance. The balance between profits and servicing customers with insurance claims is dynamic. A simplified FMEA helps keep people attuned to the risks associated with any actions under consideration.

A simplified FMEA is a method for reviewing things that can go wrong even if a proposed project, task, or modification is completed as expected. Often a project generates so much support and enthusiasm that it lacks a healthy number of skeptics, especially with regard to any effects that the project may have on things that are not directly related to it. Everyone is working on the details of getting the project going, and little effort is being spent on looking at the ramifications beyond the specific task!

The simplified FMEA form is a way of taking a critical look at a project before it is implemented; it often saves a lot of cost and embarrassment. In doing a simplified FMEA, it is assumed that all inherent components of the direct project will be done correctly. (They should have been covered in regular project reviews.) The emphasis in a simplified FMEA is on identifying affected components or issues downstream or in tangentially related processes in which issues may arise because of the project.

Just as in the simplified QFD, the critical element is getting together everyone who has anything to do with the project, especially those who have to deal with the effects of the project. These people could be machine operators, customers, or even suppliers. The proper group of participants will vary for each project.

SIMPLIFIED FMEA INSTRUCTIONS

The left side of the simplified FMEA form (see Exhibit 4-1) is a list of things that could possibly go wrong, assuming that the project is completed as planned. The first task of the meeting is to generate this list of concerns.

Exhibit 4-1. Simplified FMEA example and form

FMEA: Magnets Holding Tooling Open Ratings: 5 = Highest 1 = Lowest (or negative number)		Mount a degausser after magnets	Mount ProxSwitch and use breakaway mounts	Use electric magnets; adjust current and turn off to clean	Check with pacemaker mfg; shield if required
Concerns	**Ratings**				
The tooling will become magnetized	4	4	0	0	0
		16	0	0	0
Caught product will hit magnets, wreck machine	5	0	3	0	0
		0	15	0	0
Magnets will get covered with metal filings	2	2	0	3	0
		4	0	6	0
One operator has a heart pacemaker	5	?	0	0	?
		?	0	0	?
Magnets will cause violent tooling movement	2	0	0	3	0
		0	0	6	0
Totals **Priorities**		?	15	12	?

On this list could be unforeseen issues involving other parts of the process, safety issues, environmental concerns, negative effects on existing similar products, or even employee problems. These will be rated in importance as follows:

- 5 means that it is a safety or critical concern.
- 4 means that it is a very important concern.
- 3 means that it is a medium concern.
- 2 means that it is a minor concern.
- 1 means that whether it is an issue is a matter for discussion.

Across the top of the simplified FMEA is a list of solutions that are already in place to address the concerns and additional solutions that have been identified during the meeting. Below each solution and opposite the concern, each response item is to be rated on how well it addresses the concern:

- 5 means that it addresses the concern completely.
- 4 means that it addresses the concern well.
- 3 means that it addresses the concern satisfactorily.
- 2 means that it addresses the concern somewhat.
- 1 means that it addresses the concern very little.
- 0 or a blank means that it does not affect the concern.
- A negative number means that the solution actually makes the concern worse.

Enter this value in the upper half of the block beneath the solution item and opposite the concern. After these ratings are complete, multiply each rating by the concern value on the left. Enter this product in the lower half of each box. Add all the values in the lower half of the boxes in each column and enter the sum in the Totals row indicated near the bottom of the form. These are then prioritized, with the solution with the highest value being the number one consideration for implementation.

As in the simplified QFD, these summations are only a point of reference. It is appropriate to reexamine the concerns and ratings.

CASE STUDY: A POTENTIALLY LIFESAVING SIMPLIFIED FMEA

A high-speed production machine was experiencing wear. This wear caused the tooling to have too much play, which allowed it to rub against the product at one specific location on the machine, causing quality issues. The cost of rebuilding the machine was very high, so the manufacturing plant wanted to find other options for solving this problem.

An engineer came up with what seemed like an ingenious solution. Powerful magnets would be mounted just outboard of the machine at the problem area, near the tooling. These magnets would hold the tooling open as it went by, eliminating the chance of the tooling rubbing against the product. This solution was especially attractive because it would be inexpensive and easy to do, and it would solve the problem completely! The initial engineering study found no "showstoppers" with regard to installing the magnets. Bench tests with actual magnets and tooling indicated that it would work extremely well.

Everyone was anxious to implement this project, since all the parts were readily available, and it would be easy to install on the machine for a test. But a requirement of the Six Sigma process was to first do a simplified FMEA to see if this solution could cause other issues. So, a group of production engineers, foremen, operators, maintenance people, and quality technicians were invited to a meeting to do the simplified FMEA.

Exhibit 4-1 shows the simplified FMEA as developed in the meeting.

Most of the concerns that surfaced had doable and effective solutions. However, the concern that one operator had a heart pacemaker was a complete surprise (note the question marks in the FMEA); no one had any idea of how the magnets would affect the pacemaker.

Upon following up with the pacemaker manufacturer, it was discovered that even representatives of the manufacturer were not sure how the powerful magnets would affect the device. They did say, however, that they had serious reservations. They didn't want to commit to a level of shielding that would be sufficient to protect the operator, as they were afraid of any resultant liability.

Other options were discussed, such as reassigning the operator to another machine, but all of those options raised other issues (such as union concerns involving the reassignment). The machine operator had to be free to access all areas of the machine, so a barrier physically isolating the area around the magnets was not an option.

At that point, the option of using magnets was abandoned, because there seemed to be no way to eliminate the possible risk to the operator with the pacemaker! No other low-cost solution was identified. The machine had to be rebuilt despite the high cost.

Without the simplified FMEA, the project would have been implemented, with some real risk that the operator could have been hurt or even lost his life.

Although this example is more dramatic than most, seldom is a simplified FMEA done without uncovering some issue that was previously unknown. Most of these issues can be resolved, and it's easier to resolve them up front than afterward! In this case study, the machine was rebuilt. This would probably have been the outcome in any case; the simplified FMEA prevented the risk, cost, and embarrassment of installing the magnets, dealing with the effects, and removing the magnets.

TRADITIONAL FMEAs

As mentioned earlier, the simplified FMEA is less complex than the traditional FMEAs that are normally taught in Six Sigma. A traditional FMEA requires the people doing the form to identify each potential failure event and then the failure mode, the consequences, the potential cause, the severity, current design controls, the likelihood of detection, the frequency, the impact, risk priority, the recommended action, and the likelihood of the action succeeding. This traditional FMEA requires multiple forms, much time, and many people. Is the extra time and effort worth it?

As with the traditional QFD, perhaps it is worthwhile on very large and complex programs. However, since the simplified FMEA takes far less time, it can be used on *every* project or change, which is unlikely to happen with a traditional FMEA. This gives the simplified FMEA a real advantage, because collateral damage can occur on *all* levels of project or change.

Both the traditional and the simplified FMEAs trigger consideration of collateral damage, so one of the two should be used. Obviously the author prefers the simplified FMEA.

WHAT WE HAVE LEARNED IN CHAPTER 4

1. A simplified FMEA is usually used in the Define or Improve step of DMAIC.
2. A great deal of effort goes into making sure that the specific details of a project, process, or service are correct. However, areas that are not inherently tied to the project are often ignored.
3. A simplified FMEA emphasizes identifying concerns in other affected areas and prioritizing potential solutions to these concerns.
4. Everyone who is affected by the proposed project, process, or service should participate in the simplified FMEA.
5. Revisit the results of the simplified FMEA several times to make sure that they truly reflect the group's intent.
6. The cost of doing a simplified FMEA will be more than offset by the costs avoided that could have resulted from the project's potential negative effects on other areas.
7. Traditional FMEAs are more complex. They may be justified on extremely large and complex programs; however, they are unlikely to be used on *every* program or change, which is the value of the simplified FMEA.

RELATED READING AND SOFTWARE

D. H. Stamatis, *Failure Mode and Effect Analysis: FMEA from Theory to Execution,* 2nd ed. (Milwaukee, WI: American Society for Quality, 2003).

Robin E. McDermott, Raymond J. Mikulak, and Michael R. Beauregard, *The Basics of FMEA* (New York: Quality Resources, 1996).

Mark J. Kiemele, Stephen R. Schmidt, and Ronald J. Berdine, *Basic Statistics: Tools for Continuous Improvement,* 4th ed. (Colorado Springs, CO: Air Academy Press, 1997).

Relex FMES/FMECA, Relex Software Corporation, Greensburg, PA; www.relexsoftware.com.

Cause-and-Effect
Fishbone Diagram

What we will learn in this chapter is that it's critical that you identify and examine all of the possible causes for a problem. This chapter explains how a cause-and-effect diagram is used.

The fishbone diagram is used primarily in the Define, Analyze, and Improve steps of the DMAIC process. It helps identify which input variables should be studied further and gives focus to the analysis.

The purpose of a fishbone diagram is to identify all the input variables that could be causing the problem that is of interest. Once we have a complete list of these input variables, we identify the critical few key process input variables (KPIVs) to measure and further investigate.

Fishbone Diagram

Manufacturing. Do a fishbone diagram to list all the important input variables related to a problem. Highlight the KPIVs for further study. This focus minimizes sample collection and data analysis.

Sales and marketing. For periods of unusually low sales, use a fishbone diagram to identify possible causes of the low sales. The KPIVs enable the identification of probable causes and often lead to possible solutions.

APPLICATIONS

Accounting and software development. Use a fishbone diagram to identify the possible causes of unusual accounting or computer issues. People working in these areas respond well to this type of analysis.

Receivables. Identify periods in which delinquent receivables are higher than normal. Then, use a fishbone diagram to try to understand the underlying causes.

Insurance. Look for periods of unusual claim frequency. Then, use a fishbone diagram to understand the underlying causes. This kind of issue usually has a large number of potential causes; the fishbone diagram enables screening them to identify the critical few.

FISHBONE DIAGRAM INSTRUCTIONS

The specific problem of interest is normally the "head" of a fishbone diagram. There are six "bones" on the fish; on these bones, we list input variables that affect the problem head.

Each bone represents a category of input variables that should be considered. Separating the input variables into six categories, each with its own characteristics, helps us make sure that no input variable is missed. The six categories are Measurements, Materials, Men, Methods, Machines, and Environment. (Some people remember these as the five Ms and one E.) The six categories are what make the fishbone diagram more effective than just simply listing the input variables.

Ideally, the input variables on a fishbone should be developed by a group of "experts" working together in one room. This enables a high degree of interaction among the experts. However, if this is not feasible, it *is* possible to carry out this process on the telephone, using a computer to send updated versions of the fishbone diagram regularly to all the people contributing. It is important for all contributors to be able to see the fishbone diagram as it evolves. This will cause everyone to constantly be triggered by the six categories. Exhibit 5-1 shows an abbreviated example of a fishbone diagram done on the problem "Shaft Diameter Error."

After they have listed all the input variables, the same experts should pick the two or three KPIVs that they feel are most likely to be the culprits. Those are highlighted in boldface and capital letters on the fishbone diagram.

Exhibit 5-1. Fishbone diagram example (input variables affecting shaft diameter error)

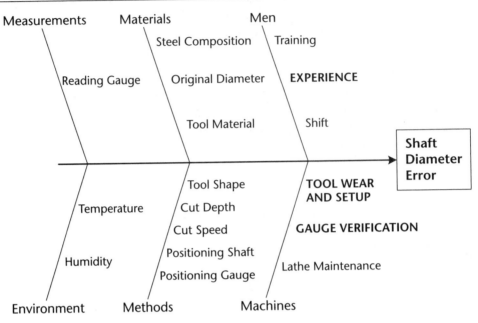

There are software packages that enable users to fill in the blanks of standardized forms for the fishbone diagram. There are also free downloads from the Internet that have forms that tie in with Excel. However, other than for the sake of neatness, preparing these diagrams by hand works just as well.

As you will see later, in Chapter 8, the fishbone diagram is the recommended tool for identifying what should be sampled in a process and what variables need to be kept in control during the sampling process. Without the kind of cause-and-effect analysis that the fishbone diagram supports, the sampling will be less focused and is more likely to be fraught with error. This is because the effort and control needed for good sampling and data collection are not trivial, so the amount of sampling must be minimized to allow everyone to get it right!

Sometimes just the process of preparing the fishbone diagram leads to the solution, because you are getting the experts together to discuss the problem, which doesn't happen without a scheduled purpose.

WHAT WE HAVE LEARNED IN CHAPTER 5

1. The fishbone diagram is used primarily in the Define, Analyze, and Improve steps of the DMAIC process.

2. Use a fishbone diagram to look for possible cause-and-effect relationships. The purpose of a fishbone diagram is to have experts identify all the input variables that could be causing the problem of interest. Once we have a complete list of the input variables, we must attempt to identify the critical few key process input variables (KPIVs) that we need to measure and further investigate.

3. The fishbone diagram is the preferred Six Sigma tool for identifying what should be sampled in a process and which variables need to be kept in control during the sampling process. Without the kind of cause-and-effect analysis that the fishbone diagram supports, the sampling would be less focused.

4. Sometimes just the process of preparing the fishbone diagram leads to the solution, because you are getting the experts together to discuss the problem with a specific focus.

5. Use of the fishbone diagram is not limited to manufacturing. Virtually any problem can be tackled using this powerful tool.

RELATED READING AND SOFTWARE

Rath & Strong Management Consultants, *Rath & Strong's Six Sigma Pocket Guide* (Lexington, MA: Rath & Strong/Aon Consulting Worldwide, 2000).

Mark J. Kiemele, Stephen R. Schmidt, and Ronald J. Berdine, *Basic Statistics: Tools for Continuous Improvement*, 4th ed. (Colorado Springs, CO: Air Academy Press, 1997).

MINITAB 13, Minitab Inc., State College, PA; www.minitab.com.

CHAPTER 6

Simplified Process Flow Diagram

What you will learn in this chapter, as in the previous chapter, is that it's critical that you identify and examine all the possible causes of a problem. In addition to the fishbone diagram discussed in the previous chapter, you can use the simplified process flow diagram to help you identify key process input variables (KPIVs) in a process. But there's a difference between these two tools.

It is important that you know where and when input variables affect a process so that you can see whether the identification of a particular variable as a factor makes sense given where the problem is being seen. Of special interest are positions in the process where inspection or quality sorting takes place or where process data are collected. By looking at data from these positions, you may see evidence of a change or a problem. By noting where different operations take place, you can also see where issues can arise. Simplified process flow diagrams are used primarily in the Define, Analyze, and Improve steps of the DMAIC process.

Simplified Process Flow Diagrams

Manufacturing. The simplified process flow diagram will focus an investigation by identifying where and when in the process KPIVs could have affected the problem. Of special interest is where data are collected during the process.

APPLICATIONS

Sales and marketing. A simplified process flow diagram will assist in identifying whether the cause of low sales is the region, personnel, or something else. This allows the team to focus on the likely area.

Accounting and software development. A simplified process flow diagram will help pinpoint the specific problem areas in a system or program. This simplifies the debugging process. Software developers are very familiar with this process.

Receivables. Identify periods during which delinquent receivables are higher than normal. A simplified process flow diagram may help in designing procedures to minimize the problem, such as discounts for early payment.

Insurance. Look for periods with an unusual frequency of claims. A simplified process flow diagram may help identify them.

A simplified process flow diagram works well when it is used in conjunction with a fishbone diagram. It can further screen the KPIVs that were identified with the fishbone diagram, minimizing the number of areas where you will have to take additional samples or data.

As for the fishbone diagram, there are software packages that enable users to fill in the blanks of standardized forms for the process flow diagram. There are also free downloads from the Internet that have forms that tie in with Excel. However, other than for the sake of neatness, preparing these forms by hand works just as well.

This chapter looks at many areas of Lean Six Sigma, where reducing lead time, reducing work in process, minimizing wasted motion, optimizing work area design, and streamlining material flow are some of the techniques used to reduce manufacturing costs. Both Lean and traditional Six Sigma use simplified process flow diagrams to assist in identifying quality and cost issues. Both Lean and traditional Six Sigma are very timely given the tight economic and competitive market companies are in.

Of special interest to Lean Six Sigma is noting where product assembly, transfer, handling, accumulation, and transport occur. Not only are these areas of potential cost savings, but in many of these areas, excess handling and inventory are problem sources and can delay timely response to an issue.

Lean manufacturing got its start at Toyota in Japan, but now U.S. companies are relooking at their processes using Lean Six Sigma. As reported in the *Wall Street Journal* ("Latest Starbucks Buzzword: 'Lean' Japanese Techniques," by Julie Jargon, August 4, 2009), Starbucks is one of these companies.

Starbucks may seem like an odd place to be using this tool, but Starbucks is very sensitive to how long a customer must wait before being served. Customers will walk out if they feel that the wait is excessive. By applying Lean techniques, according to the article, even a well-run Starbucks store was able to cut its average wait time. The position and means of storage of every element used to make drinks was analyzed to minimize wasted motion. The position and storage of all the food items was similarly examined. In this way, service time was reduced, along with wait time.

The traditional process flow diagram for Six Sigma shows the steps in the process, with no relative positions or time between these steps. In doing a simplified process flow diagram for Lean Six Sigma, you will show the relative distance and time between process steps, and, if desired, numerically show actual distances, time intervals for product flow, and inventory builds. The intent is to identify areas where handling and process time can be reduced. In Lean Six Sigma, you want to be able to separate meaningful work from wasted time and motion. This wasted time and motion could include searching for components, which perhaps would be easier to access if they were in color-coded bins. Ideally, a person's efforts would stay the same, or be reduced, while more product is made or services performed as a result of improved process flow and efficiency. Sometimes, a person's or product's movements are so elaborate that the map that is developed is called a "spaghetti map" or a "value stream map." But, no matter what it is called, it is still a process flow diagram.

SIMPLIFIED PROCESS FLOW DIAGRAM INSTRUCTIONS

A process flow diagram shows the relationships among the steps in a process, or the components in a system, with arrows connecting all the pieces and showing the sequence of activities. Some texts and software for traditional process flow diagrams use additional geometrical shapes—circles, inverted triangles, and so on—to differentiate among types of functions in the process. I choose to keep it simple. Exhibit 6-1 shows a simplified process flow diagram.

Like the fishbone diagram, the simplified process flow diagram is not limited to solving problems in a manufacturing process. Also, a simplified process flow diagram is not limited to a physical flow map. The flow could be related to time or to process steps, not just to place.

Exhibit 6-1. Simplified process flow diagram, shaft machining

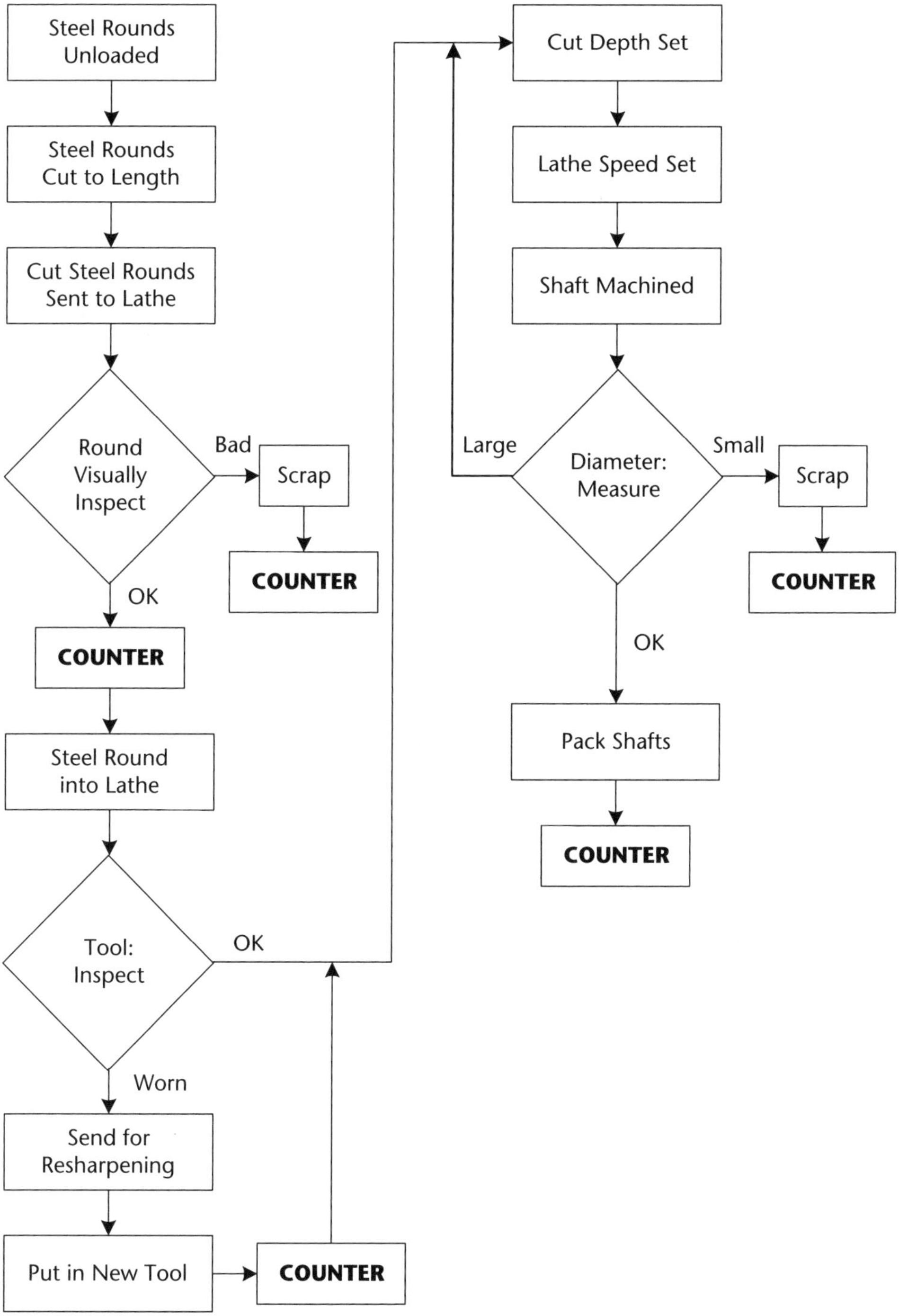

Finally, process flow diagrams are not just for solving problems. They can also be used to configure a proposed new process.

CASE STUDY: EXCESS MATERIAL HANDLING

In a Lean Six Sigma project, process flowcharts showed that, in a manufacturing plant that ran 24 hours per day, the product flow for the day shift was different from that for the other shifts. In trying to understand why, it was found that the quality engineer insisted on being the final decision maker on the resolution of any product that was put into quality hold by the on-line quality checks. On day-shift production, the quality engineer went to the held product and made timely decisions based on the inspection data. This caused the product to be either reinspected by the day shift or sent to the warehouse. On the other shifts, the held product was taken to the warehouse until the next day shift, when the quality engineer would review the data. Any product that the engineer deemed bad enough to require reinspection then had to be removed from the warehouse and taken to an inspection station. This led to extra handling of the off-shift product that required reinspection.

Once this practice was brought to the attention of management, the quality engineer was asked to document how he made the decision to reinspect or ship, and the shift foremen were then trained in this task. Also, the day-shift inspectors were distributed to the off-shifts so that any held product could be reinspected in a timely manner.

This change reduced material handling, enabling the reduction of one material handler on the day shift. But even more important, the reinspection on the off-shifts gave the shifts quicker feedback on any quality issues they had. This helped improve the off-shifts' overall product quality (reduced the variation), reducing the quantity of held product requiring reinspection. This improved the overall profitability of the plant.

An issue to keep in mind when implementing Lean Six Sigma is the legacy of scientific management, or Taylorism, which was developed in the late nineteenth and early twentieth centuries. Taylorism emphasized the careful study of an individual at work to develop precise work procedures that were broken down into discrete motions to increase efficiency and to decrease waste. Over the years, this process has at times included time and motion studies and other methods to improve managerial control over employees' work practices. However, these scientific management methods have sometimes led to ill feelings on the part of employees, who felt that they had been turned into automatons whose jobs had been made rote with little intellectual stimulation. Casual work attitudes on some assembly lines are one manifestation of this issue. Although Lean Six Sigma implementers will be quick to point out how what they do is different from those earlier approaches to efficiency, there are workers who are suspicious of any attempt to optimize work practices.

Probably the most effective way to minimize conflict during Lean Six Sigma projects is to make sure that communication with everyone involved is frank and open, with the emphasis being on eliminating boring and wasted tasks that do not add to job satisfaction. These improvements generally lead to improved productivity and quality without generating employee animosity. Also, when making an improvement to a work center that includes optimizing parts placement, it is wise to be sure that an FMEA is done before the actual change is made to reduce unforeseen problems.

WHAT WE HAVE LEARNED IN CHAPTER 6

1. Simplified process flow diagrams are used primarily in the Define, Analyze, and Improve steps of the DMAIC process.
2. A simplified process flow diagram will help further pinpoint the areas in which efforts should be concentrated. Of special interest in a process is the areas where data are collected, since this will often help focus the study to a defined process area.
3. Simplified process flow diagrams work well in conjunction with a fishbone diagram. By using the two, the areas that need to be further investigated and sampled are minimized.
4. A simplified process flow diagram is not limited to a physical flow map. The flow could also be related to time or to process steps.
5. Besides being used for problem solving, process flow diagrams can be used to configure a proposed new process.

RELATED READING AND SOFTWARE

Rath & Strong Management Consultants, *Rath & Strong's Six Sigma Pocket Guide* (Lexington, MA: Rath & Strong/Aon Consulting Worldwide, 2000).

Mark J. Kiemele, Stephen R. Schmidt, and Ronald J. Berdine, *Basic Statistics: Tools for Continuous Improvement*, 4th ed. (Colorado Springs, CO: Air Academy Press, 1997).

MINITAB 13, Minitab Inc., State College, PA; www.minitab.com.

CHAPTER 7

Correlation Tests

W hat you will learn in this chapter is how to discover the key process input variables (KPIVs) that may have caused a change in a process or product. To find them, we will be doing correlation tests.

In some Six Sigma classes, regression analysis is used to find correlations. A mathematical curve is fit to a set of data, and various techniques are used to measure how well the data fit this curve. The curve is then used to test for correlations.

These methods require a high degree of skill and generally are not friendly to those who are not doing this kind of analysis almost daily. Thankfully, most Six Sigma work can be done using the tools we have already covered, as long as we are willing to do some visual examination of data and their related graphs. Correlation tests are used primarily in the Define, Analyze, and Improve steps of the DMAIC process.

Something in a process or product has changed, and we would like to discover the KPIV(s) that caused the change. Time and position are the critical factors in doing the analysis.

Correlation Tests

Manufacturing. Do a time plot showing when a problem first appeared or when it comes and goes. Do similar time plots of the KPIVs to see if a change in any of these variables coincides with the timing of the problem change. If one does, do a controlled test to establish the cause-and-effect relationship for that KPIV.

APPLICATIONS

Sales and marketing. For periods of unusually low sales activity, do a time plot showing when the low sales periods started and stopped. Do similar time plots of the KPIVs to see if a change in any these variables coincides with the low sales period. If one does, do a controlled test to establish the cause-and-effect relationship for that KPIV.

Accounting and software development. Do a time plot of unusual accounting or computer issues. Do similar time plots of the KPIVs to see if a change in any these variables coincides with the issues. If one does, do a controlled test to establish the cause-and-effect relationship for that KPIV. People in these areas respond well to this type of analysis.

Receivables, insurance, and other such areas. Identify periods when delinquent receivables are higher than normal or when the frequency of claims is unusual. Then do a time plot of the problem and the related KPIVs. For any variable that shows coincident change, check for cause-and-effect relationships with controlled tests.

CORRELATION TEST INSTRUCTIONS

We first isolate when and where the problem change took place. We do this by preparing a time plot or a position plot of every measurement of the process or product we have that indicates the change. From these plots, we can often define the time and/or the position of the change within a very narrow range. If the change indicated by the plot is large compared with other data changes before and after the incidence and the timing corresponds to the recognition of the observed problem, it is generally worthwhile to check for correlations.

The next thing to do is to look for correlations with input variables, often using graphs of historical data. If we don't know the KPIVs, we must use a fishbone diagram or a process flow diagram to identify them. We do time plots or position plots of every KPIV, focusing on the previously defined time period or position. Any input variable that changed at nearly the same time or position as the problem is suspect.

When we find multiple time or position agreements of change between the problem and an input variable, then we must do controlled tests in which we control everything but the suspicious variable. In this way, we can establish cause-and-effect relationships.

If more than one KPIV changed, there could be an interaction between these variables, but usually one KPIV will stand out. Looking at extended time periods will often rule out input variables that do not correlate consistently.

Later in the text, we will learn numerical methods of testing for statistically significant changes. These tests can be used to test for significant changes in the data from immediately before and immediately after the problem begins. They can be used on both the problem data and the KPIV data. However, if there is multiple time agreement on changes between the problem and the KPIV, these extra tests are often not needed. In any case, we will have to run controlled tests to prove cause-and-effect relationships.

Exhibit 7-1 shows simplified plots of a problem and the process KPIVs (A, B, and C). You can see that this visual check is very easy, and the cause of the problem is often obvious once plots of the problem and the KPIVs are compared with each other.

Exhibit 7-1. Correlation illustration plot

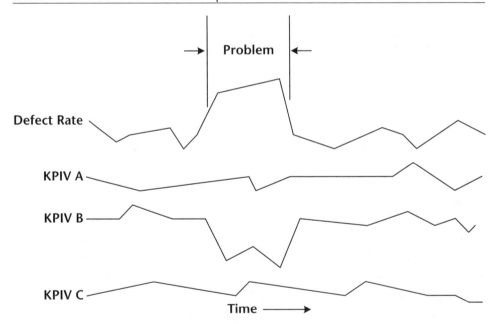

KPIV B certainly looks suspicious, given that it had a change in the same time interval as the problem, with both the beginning and the end of the time period matching. As a first test, I would expand the time of the data for both the process defect rate and KPIV B to see if this change agreement is truly as unique and correlated as it appears in these limited data. Remember, however, that this test will never be definitive. It will only hint at the cause. A controlled test that holds all possible variables (except for KPIV B in Exhibit 7-1) the same will be required. We would intentionally change KPIV B in accordance with the plot in Exhibit 7-1 and see if the problem responds similarly. Only then have we established a cause-and-effect relationship.

When time plots of variables are compared with the change we are studying, it is important that any inherent time shift be incorporated. For example, if a raw material is put into a storage bin with the inventory from three days previously, this three-day delay must be incorporated when we are looking for a correlation between that raw material and the process.

Showing Cause and Effect

TIP

Correlation doesn't prove cause and effect. It just shows that two or more things happened to change at the same time or at the same position. There have been infamous correlations (for example, stork sightings versus birth rates) that are just coincidental or that have other explanations.

To show cause and effect, you must run controlled tests in which only the key test input variable is changed and its effect is measured. Normally, historical data can't be used to prove cause and effect because the data are too "noisy" and the other variables are not being controlled.

CASE STUDY: INCORRECT CAUSE

Glass containers were heated to very high temperatures in an indexing device, and it was critical that the containers softened at the same position on this indexing heating machine. This was critical because several processes on this machine relied on the glass having certain viscosities at different stations on the machine. The people running this indexing heating device had historically complained that the glass containers softened at different positions on the machine, causing issues with the processes that occurred at the different machine stations and therefore hurting the quality of the final product.

It was believed that the cause of this problem was large variation (high sigma) within the glass wall of each container. This was not a new problem, and over the years the tolerances on the container wall variation had been tightened. These tolerances were now so tight that the container plant was incurring large losses trying to meet these tight specifications. Because the complaints continued, however, a major project was started to further reduce the wall variation within each container.

To find out how critical the variation in wall thickness was, the project team ran a large group of containers with great variation in wall thicknesses on the indexing heating machine and compared the results against those obtained by running a group of containers with little variation in wall thicknesses. Using the statistical tests that will be covered later, the project team found *no* statistically significant difference between the two groups.

The historical belief that the wall thickness sigma within each container was causing the container to soften at different positions was wrong!

A search was started to find the KPIVs that influenced the position at which the glass softened. Looking at periods of time when the complaints were highest versus times when the complaints were reduced, one of the KPIVs that was found to correlate was the *average* wall thickness of each container—not the wall *variation* within each container. When a test was run with containers grouped with others having similar average wall thicknesses, each container softened consistently with others in the group, at different positions on the machine. Again, the variation (sigma) within each individual wall had no effect. The test was repeated in different plants on similar machines with the other variables in control, and the results supported the cause and effect.

This subtle finding that the average wall thickness was the KPIV and that variation within each container wall did not correlate changed the way the container was manufactured. This saved the container manufacturer $400,000 per year, because it had been scrapping containers with large variations within the wall. It also saved the plants running the heating machines $700,000 per year through better yields.

The tool that triggered the realization that container wall *variation* was not the culprit was visually checking for correlations in plotted data on wall variation versus softening position on the indexing machine. No visual correlation was seen. However, there *was* a visual correlation between *average* wall thickness change and softening position.

These initial observations were followed up with quantitative statistical tests that checked for significant change, but the correlation tests done on visual plots were the breakthrough trigger. And, of course, controlled tests were needed to prove cause and effect.

WHAT WE HAVE LEARNED IN CHAPTER 7

1. Correlation tests are used primarily in the Define, Analyze, and Improve steps of the DMAIC process.
2. In some Six Sigma classes, regression analysis is used to find correlations. These methods require a high degree of skill and generally are not friendly to those who do not do this kind of analysis almost daily.
3. When something has changed in a process or product, we would like to discover the key process input variables (KPIVs) that caused the change. Time and position are the critical factors in doing the analysis. Using data plots, we first isolate when and where the problem change took place.
4. Look for a matching time period or position change on the data plots of all input variables. Identify all KPIVs that have a change that correlates with the problem.
5. Test for a cause-and-effect relationship by running controlled tests with only the suspect KPIV being changed.
6. Statistical tests for significance, which will be covered later in the text, can also assist in checking for correlations. But often the visual correlation using data plots is sufficient, especially when you see multiple correlations of timing between the problem and a KPIV.

RELATED READING AND SOFTWARE

Rath & Strong Management Consultants, *Rath & Strong's Six Sigma Pocket Guide* (Lexington, MA: Rath & Strong/Aon Consulting Worldwide, 2000).

Mark J. Kiemele, Stephen R. Schmidt, and Ronald J. Berdine, *Basic Statistics: Tools for Continuous Improvement*, 4th ed. (Colorado Springs, CO: Air Academy Press, 1997).

MINITAB 13, Minitab Inc., State College, PA; www.minitab.com.

PART III

**Foundations for Using
Statistical Six Sigma Tools**

CHAPTER 8

Getting Good Samples and Data

What you will learn in this chapter is how to take good samples and get good data. Without these, even the best thinking won't matter much because you won't be able to put it to the test properly.

In later chapters, you will see how to calculate the minimum sample size and how to verify that the gauge that is being used to measure a product is giving you data that are sufficient for your needs. Just as important, however, is making sure that your sample and your data truly represent the population of the process that you wish to measure. The whole intent of sampling is to be able to analyze a process or a population and get valid results without measuring every part or every component, so sampling details are extremely important.

Issues in Getting Good Data

Manufacturing. Samples and the resulting data have to represent the total population, yet the processes controlling the population are often changing dramatically because of changes in people, shifts, environment, equipment, and other such factors.

Sales. Sales forecasts often use sampling techniques in making their predictions. However, the total market may have many diverse groups to sample. These groups may be affected by many external drivers, like the economy.

APPLICATIONS

Marketing. What data should be used to judge a marketing campaign's effectiveness, since so many other factors are changing at the same time?

Software development. What are the main causes of software crashes, and how would you get data to measure the crash resistance of competing software?

Receivables. How would you get good data on the effectiveness of a program intended to reduce overdue receivables, given that factors like the economy exert a strong influence and change frequently?

Insurance. How can data measuring people's satisfaction with different insurance programs be compared when the people covered by the programs are not identical?

We have all seen the problems that pollsters have had in predicting election outcomes based on sampling. In general, the problem has *not* been in the statistical analysis or in the sample size. The problem has been picking a group of people to sample who truly represent the electorate!

The problem of sampling and getting good data has several key components. First, the people and the methods used for taking the samples and data affect the randomness and accuracy of both. Second, the product population is diverse and often changes, sometimes quite radically. These changes occur over time and can be affected by location. To get a true reflection of a population, anyone who is sampling and using data must be aware of all these variables and somehow get valid data despite them.

I will share some of the difficulties and challenges that my teams have experienced in getting representative samples and data. I will then discuss some approaches for getting useful and valid data despite these issues. Most of the examples pertain to manufacturing, but I will explain later how the approach recommended for getting good data applies to many other applications.

HAWTHORNE EFFECT

As soon as anyone goes out to measure a process, things change. Everyone pays more attention. The process operator is more likely to monitor his process closely, and quality inspectors are likely to be more effective in segregating defects. The result is that the product you are sampling is not likely to represent that produced by a normal process. This is true even when people are polled on an issue, in that the answer may be the result of far more careful thought than the impulses or knee-jerk reactions that might guide those people in their daily actions.

There have been many studies done on how people react to having someone pay attention to them. Perhaps the most famous is the Hawthorne Study, which was done at a large Bell Western manufacturing facility—the Hawthorne Works—in Cicero, Illinois, from 1927 to 1932. This study showed that any gain realized during a controlled test often results from the positive interaction between the people doing the test and the participants, and also from the interaction among the participants. The people might begin to work together as a team to get positive results. The actual variable change being tested was often not the driver of any improvement.

One of the tests at the Hawthorne facility involved increasing the light level to study the influence of the increased light on productivity. Productivity did indeed increase when the light level was increased. However, in a control group for which the light level was *not* changed, productivity also improved by the same amount. It was apparently the attention given to both groups that was the positive influence, not the light level. In fact, when the lighting was restored to its previous level, the improvement in productivity continued for some period of time. This effect of attention has become known as the Hawthorne Effect.

Any data you take that show an improvement that you think is the result of a change you have implemented must be suspect because of the Hawthorne Effect. Your best protection against making an incorrect assumption about improvement is to take data simultaneously from a parallel line with an identical process, but without the change (a control group). However, the people in both groups should have had the same attention paid to them, attended the same meetings, and so on. An alternative method is to collect line samples just before the change is implemented, but after all the meetings, interaction, and other such actions have taken place. These "before" samples would be compared with the "after" samples, with the assumption that any Hawthorne Effect is included in both.

There was a different result, however, in another study in the same Hawthorne facility. In this case, the participants in the study were afraid that the test results were going to negatively affect their jobs, and, as a group, they had agreed that their productivity would *not* improve, no matter what changes were implemented, so of course it didn't. Doing a test in this environment would make it very difficult to ascertain whether a change was good or bad, since the experiment could be undermined. In this kind of environment, the only way to get good data is to do a surreptitious change, unless the change is so basic to the process that its results can't be denied.

If you ask an inspector to pick up and inspect a product at random, there is a good chance that the sample will be biased toward any product with a visible defect. This is because inspectors are accustomed to looking for defects and because they believe you are there because of problems with defects, so they want to be helpful.

I once ran a test where product was being inspected on the line, paced by the conveyor speed. I collected the rejected product and isolated the packed "good" product from this same time period. Without telling the inspectors, I then mixed the defective product back with the "good" packed product. Without telling the inspectors that they had already inspected the product, I had the same inspectors inspect this remixed product off the line, where the inspectors weren't machine-paced. The defect rate almost doubled. (Interestingly, the customer had not been complaining about the product coming through the on-line inspection.)

When the product was inspected without time constraints, the quality criteria apparently tightened, even though no one had triggered a change in the criteria. Or maybe the inspectors just became more effective. Another possibility is that the inspectors felt that I was checking on their effectiveness in finding all the defects, so they were being extra conservative in interpreting the criteria. In any case, someone using data from the off-line inspection would get a defect rate almost double the rate that was seen with the on-line inspection, from an equivalent production process. Therefore, if someone had implemented a change and was checking its effectiveness by checking for defects off-line, the change would have had to reduce the actual defects by half to look even equivalent to the historical data from on-line inspection. Obviously this would be problematic.

Time considerations are not the only influence on quality criteria interpretation. To check the optics on a parabolic reflector, an inspector would insert the reflector into a fixture that seated the reflector precisely over a light source. The inspector would then make a judgment on the quality of the resultant projected image. Too many "poor" readings would cause the product to be scrapped and the reflector-forming process to be reset.

As a test, on a day when there was an unusually high incidence of "good" optical readings, I collected the relatively few reflectors that had "poor" readings. On a later date, when the process was generating a lot of "poor" optical readings, I reintroduced the reflectors that had earlier been judged "poor." They were now judged as "good." Because of the qualitative nature of the criteria, the judgment of "good" or "poor" apparently became relative to the average optics that the inspector was currently seeing.

Sometimes people become very defensive (or maybe even offensive) when samples are taken from their process. In one of the case studies I relate later, employees of a manufacturing plant thought that its defects were caused by bad raw materials. When a team began collecting defects on one of the plant's production lines and correlating them with specific problems on that line, the line operator grieved to his union that he was being harassed, since the engineering team was not looking at the raw materials, which the operator was *sure* were causing the problem. (Incidentally, the problems did prove to be related to the line and were not caused by raw materials.)

Case Study: Outcome Bias

An engineer was automating a plant's production line with some in-line automatic inspection equipment. To validate the equipment, one of the Six Sigma test requirements was to see whether there was a statistically significant difference between the products inspected by the automatic equipment and the products inspected by the people on-line.

The first test results showed that the packed products that had been inspected by the automatic inspection equipment had a significantly higher defect level than the products inspected by the people, so the automatic inspection equipment failed the test. This result surprised the engineer, because on previous tests with "master" defects, the automatic equipment had appeared to be very good.

The tests comparing the automatic equipment and the manual on-line inspection involved comparing alternate intervals of products inspected only by the equipment and products inspected only by the people. By doing this, the engineer felt that he was removing any variable related to the overall incoming quality of the product.

After the equipment failed this initial test, the engineer reviewed the sampling technique and the manner in which the two samples were compared and realized that his methodology could be flawed. First, the people who were inspecting the product on-line did not want the automatic inspection equipment to work, since they felt that it might jeopardize their jobs. The engineer suspected that the inspectors were being extra conservative in manually inspecting the product during the test. Second, the off-line people who were reinspecting the samples of product also felt that their jobs could possibly be at risk, since their jobs were also classified as "inspector." The engineer suspected some bias in their judgment of what they were calling a "defect" when reinspecting both groups of product. Third, the quality manager, who from the beginning had proclaimed that the automatic inspection equipment could never have the diverse inspection ability of a human, may have had some bias in any data outcome analysis.

To correct for these sampling and comparison deficiencies, the engineer running the tests changed his sampling and reinspection procedures. He decided to gather his samples on random shifts over a week's time period. Without prior notice, he would go out to the production line and collect samples of packed product that had just been inspected by the people. He would then have them stop manually inspecting, and he would start the automatic inspection equipment. Then he again collected packed product samples. Finally, he would turn off the inspection equipment and resume the standard manual inspection. He did not collect samples again; his samples were always taken just before and during automatic inspection, with no prior warning. These two groups of samples were numbered using a random number generator, so that only he knew which of the samples came from the automatic inspection and which came from the on-line people. After collecting samples randomly over a week, he gave the samples to the quality department to inspect. He did not give the secret code for identifying the inspection method to the people who were reinspecting the products or to the quality manager. (Incidentally, this approach did not go over well with the quality manager. He put in a strong protest to the plant manager.)

Only after all the data on the reinspected product were published was the code identifying the inspection method released. Then the statistical tests for change, which are covered later in the text, were applied. There was no statistically significant difference between the two groups of reinspected product. The automatic inspection equipment passed its Six Sigma test requirements.

VARIABLES

There are many factors that can affect sampling. Here are a few to make you more aware of the complexity and difficulty of getting good data.

Sometimes an inspector "adjusts" data, truly believing that the adjustment gives a truer picture of the actual process. Here's an example.

I was watching an inspector who was inspecting product on a high-speed production line. On a regular basis, a random product was picked from the conveyor line and placed onto a fixture that had several electronic gauges that took key measurements. These measurements were displayed on a computer screen and then automatically sent to the quality database, unless the inspector overrode the sending of the data. The override was intended to be used only if the inspector saw a very specific problem, like the product not being seated properly in the fixture, that would necessitate a new reading on that product.

As I was observing, I saw the inspector periodically override the sending of the data, even though I saw no apparent problem with the seating. When I asked her why the data were not being sent, she replied that the readings looked unusual and that she didn't think they were representative of the majority of the products being measured. She didn't want what she thought were erroneous data being sent to the system, so she overrode the sending of the data and went on to the next product. She didn't even reread the product.

She proudly told me that she had been doing this for years and that she had trained other inspectors to do the same. So much for using *those* data! Anyone who ran a test on this line, taking his own quality samples, would probably find more variation in the self-inspected samples than the historical data in the quality system would show.

Getting random samples from a conveyor belt is not always easy. Sometimes the production equipment has multiple heads that unload onto a conveyor belt in a nonrandom fashion. Some of the stations on the production machine may send all of their products down one side of the conveyor belt, so that someone taking samples from the other side of the conveyor belt may never get any product from some of the production stations.

The start-up of any piece of equipment often generates an unusually high incidence of defects until the equipment is debugged. After a shift change in multiple-shift plants, it may take some time for the new operator to get the machine running to her own parameters, during which time the quality may suffer. Absenteeism and vacations cause less experienced people to operate equipment, with generally lower quality. Maintenance schedules can often be sensed in product quality. Another influence on quality is scheduling—which production line is scheduled on which product. And, of course, there are variations in humidity, temperature, and other such conditions.

Certainly the overall quality is affected by these variables and more. In fact, a case could be made that many of the quality issues come from these "exception" mini-populations. So, how can you possibly sample in such a way that all these variables are taken into account?

First, you probably can't take samples that will account for all of the possible combinations just listed. In fact, before you begin to take any samples, you have to go back to the first step in the DMAIC process and define the problem. Only with this better definition of the problem will you be able to ascertain what to sample.

PROCESS OFF-CENTER

Is the problem that the process is off-center? For example, are you worried about machined shaft diameters when the initial data indicate that, on average, they are running too large? Is the problem that all order takers are consistently making too many errors? Are almost all orders for a product being filled late? If a problem is of this nature, then that problem is perhaps best addressed by improving the whole process, not by focusing on the variation caused by the exception mini-populations. If this is the case, it makes collecting samples and data and measuring change a lot easier than if you had to gather samples and data on each peculiar part of the population.

When you are attempting to measure your success in centering a process or changing the process average, you want to collect samples or use data that represent a "normal" process, both before and after any process adjustment. You don't want samples from any of the temporary mini-populations.

One of the ways to identify the "normal" population is to do a fishbone diagram where the head of the fish is nonnormal populations. When you do this, the bones of the fish (the variables) will be all the variables that cause a population to be other than normal. You will then make sure that the time period during which you collect samples and data is free from any of the conditions listed on the fishbone.

Let's look at an example. Let's say the problem is the issue mentioned earlier, that machined shafts are generally running too large. Previously we did a fishbone diagram on shaft diameter error. Let's look at that fishbone again (Exhibit 8-1) in reference to this example.

Let's look at the key process input variables (KPIVs) shown on this fishbone to determine which one(s) would be likely to cause the shaft diameters to run off-center, generally too large. The expert-picked KPIVs are experience of the operator, tool wear and setup, and gauge verification.

Exhibit 8-1. Fishbone diagram of input variables affecting shaft diameter error

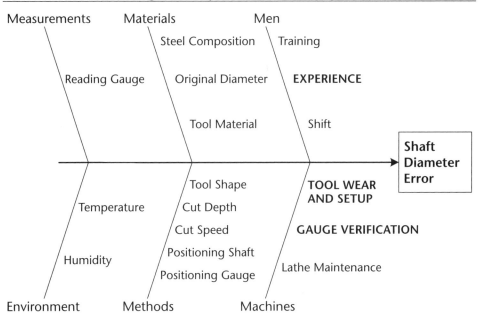

The experience of the operator would perhaps cause this problem for short periods, but the problem would not be ongoing; we would expect to have an experienced operator at times. Gauge setup and verification could account for the problem, since the gauge could be reading off-center such that the diameters would be generally too large. However, for this example, let's assume that we check, and we are satisfied that the simplified gauge verifications (which will be covered in detail in the next chapter) have been done on schedule and correctly.

That leaves tool wear and setup. Tool wear would most likely cause the diameters to change as the tool wears, but then the cycle would start over when the tool was changed. However, if the tool setup is incorrect, it could position the tool incorrectly all the time. This could conceivably cause the diameters to be generally high. So, this is the variable we want to test.

We will want to run a test to see whether having the operator set up the tool with a different nominal setting will make the process more on-center. We want to do random sampling during the process, with the tool setup being the only variable that we change.

Since the effect of tool setup is what we want to measure, we want to control all the other input variables. In accordance with the fishbone diagram in Exhibit 8-1, we especially want to have experienced people working on the process, and we want to be sure that the simplified gauge verification was done. These were the input variables that had been defined by the experts as being critical. We will also use a tool with "average" wear, so that tool wear is not an issue. The test length for getting samples will be short enough that any additional tool wear during the test will be negligible.

We will use an experienced crew on the day shift, verifying that the shaft material is correct and that the lathe is set up correctly (cutter depth, cutter speed, and position of the shaft and gauge). We will make sure that lathe maintenance has been done and that the person doing the measurements is experienced. We will minimize the effects of temperature and humidity by taking samples and data on the "normal" process, then immediately doing the test with the revised setup and taking the test samples and data. We will take samples and data only during these control periods.

Note that we used the fishbone both to show the KPIVs and to help us pick the input variables that we logically concluded could be causing the issue. Without this process of elimination, we would have had to test many more times. By limiting and controlling our tests, we can concentrate on getting the other variables under control, at least as much as possible.

If this test on tool setup did not solve the problem of diameters being too large, we would then go back and review our logic, perhaps picking another variable to test.

Getting Good Samples and Data

TIP

Use good problem definition, a fishbone diagram, and any of the other qualitative tools to minimize the number of variables you have to test. Then do a good job of controlling the other variables during the test.

Sample sizes, needed statistical analysis, and other such material will be covered in later chapters. In this chapter, we are emphasizing only the nonnumerical problems related to getting good data.

The example I just gave pertained to manufacturing. But what if the problem is in an office, like the earlier-mentioned issue of almost all order takers making too many errors? Again, we would use a trusty fishbone diagram (Exhibit 8-2), with the head being "order error rate." Let's assume that Exhibit 8-2 shows the fishbone completed by a group of "experts." These experts could have included experienced order takers, their managers, employees who pack the orders based on the forms, billing staff, customer service personnel, and the customers.

Exhibit 8-2. Fishbone diagram of input variables affecting order error rate

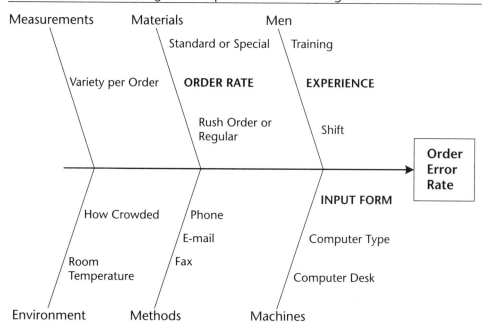

The KPIVs picked by the experts are order rate, experience, and input form. Let's see which of these KPIVs make sense as being the cause of the problem as defined.

The order rate would vary, with some time periods having a low order rate. This isn't consistent with our problem definition that the error rate is *consistently* too high, so order rate is not the variable that we will test initially. Experience presumably varies among the order takers, so again that is not consistent with the problem definition that almost all the order takers were making too many errors. Only the input form looks as if it would affect most order takers consistently.

Thus, we want to test whether a redesign of the form that the order takers use can minimize these errors. Just as we did in the manufacturing example, we will want to control all of the variables except the one we wish to test and to collect our samples and data only during these controlled periods. Of special concern as factors to control are the highlighted KPIVs, since these are the variables that the experts identified as being most likely to affect the order error rate.

Therefore, we will review only orders taken by experienced order takers during periods of time when the input of orders is at a somewhat average rate (that is, not exceptionally high or exceptionally low). We will do this on the day shift, making sure that the room temperature and the number of people in the area are pretty much normal. To take out the effect of the different methods, we will evaluate only orders taken by phone. The base sample data to get the normal error rate will be taken on one day; the test samples with the new form will be taken the following day.

CENTERING OR VARIATION?

The examples just given were a way to get good samples and data when the problem definition indicated that the problem was related to a process that was not centered, so our emphasis was on improving the total process. As you will see later in the book, centering a process, or moving its average, is generally much easier than reducing its variation. Reducing process variation often involves a complete change in the process, not just a minor adjustment.

If the problem definition indicates that the problem is the variation and that the centering of the process is not the issue, then you have no choice but to try to identify the individual causes of the variation and try to reduce their effect.

Again, you will save yourself a lot of trouble if you can make the problem definition more specific than just stating that the variation is too high. Does the problem happen regularly or at a spaced frequency? Is it related to shift, machine, product, operator, or day? Any specific information you can find will dramatically reduce the number of different mini-populations from which you will have to gather samples and data. This more specific problem definition can then be compared with the related fishbone diagram to try to isolate the conditions that you must sample.

PROCESS WITH TOO MUCH VARIATION

Suppose our earlier shaft diameter error problem had been defined as being periodic, affecting one machine at a time, and not being an off-center process problem. Let's revisit the fishbone diagram with this new problem definition in mind (see Exhibit 8-3).

Since the problem was defined as being periodic, let's see which of these input variables would be likely to be associated with a production line time period related to the problem. It appears that each KPIV (experience, tool wear and setup, gauge verification, and lathe maintenance) may have different periods. With this insight, we need to go back and see

Exhibit 8-3. Fishbone diagram of input variables affecting shaft diameter periodic variation error

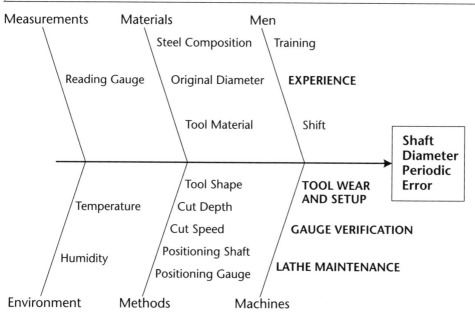

whether it is possible to get even better problem definition that will allow us to focus more sharply on the specific problem.

Assume that we go back to the customer or whoever triggered the issue and find that the problem occurs every several weeks on each line, but not on all lines at the same time. Let's look at our KPIVs with this in mind. Experience would be random, not every several weeks. The cutting tools are replaced every several days, so the time period doesn't match. Simplified gauge verifications are done monthly, so that cycle also doesn't fit. However, lathe maintenance is done on a two-week cycle, one machine at a time. This variable fits the problem definition.

We want to control everything other than maintenance during our sample and data collection. We will change the cutting tool frequently, verifying its setup, to make sure that this is not an issue. We will have experienced people working on the process, and we will be sure that simplified gauge verification was done. All of these input variables had been defined by the experts as being critical, so we want to be sure to have them in control.

We will use an experienced crew on the day shift, verifying that the shaft material is correct and that the lathe is set up correctly (cutter depth, cutter speed, and position of the shaft and gauge), and that an experienced person will be doing the measurements. We will minimize the effects of temperature and humidity by taking samples and data at the same time each day. Since we don't know whether the problem is caused by not doing maintenance often enough or whether the lathe takes some time to debug after maintenance, we will probably want to take samples daily for at least two weeks to get a better idea of the actual cause.

As you can see in all the examples given here, good problem definition combined with a fishbone diagram will focus us on which samples and data we need. The detail within the fishbone will further help us make sure that the other variables in the process being measured are as stable as possible, with the exception of the variable being evaluated.

Once a change has been implemented, samples and data must be collected to verify that the improvement actually took place. The same care must be taken in collecting samples and data validating the improvement as was taken during the problem-solving process. You can use quality department data for validation if the means of collecting data stays consistent before and after the change. If you can't trust the quality department data, you will have to use data samples.

IMPORTANCE OF GETTING GOOD SAMPLES AND DATA

Minimum sample sizes, simplified gauge verification, and statistical tests to validate significant change all play a part. But the sample and data collection must be right to start with, so that you have data to analyze that are truly representative of the process you are checking.

While you are testing, collecting data, and validating the process improvement, you and others must be alert for any event that makes your conclusions suspicious. When in doubt, don't believe the results. Redo the test!

The use of good problem definition and the fishbone diagram to help decide what to sample is valid for many applications.

Sales managers, store salespeople, distribution center employees, and others can assist in doing a fishbone diagram when the head of the fish is "inaccurate sales forecasts." Just as in manufacturing, there are probably many influencing variables. The problem definition and fishbone will help in deciding on the critical variables and in making sure that your sampling and data are focused and are minimally affected by input variables other than the one you are testing.

Marketing folks can get advertising experts from newspapers, TV, and magazines to assist in doing a fishbone when "ineffective advertising" is the head of the fish. Variables may be advertising style, media type, frequency, and market, among others. This information helps determine what and how to sample.

Software developers can get users to help with the fishbone when the head is "software crashes." Output can be used to identify which areas to focus on to get data on each key cause of software crashes.

Everyone from sales to accounting can contribute to a fishbone with "too many overdue receivables" as the head. Again, this is needed before determining what to sample to get good data on the problem.

Suppose an insurance company feels that it has too many policies with only minor differences. A group of salespeople and customers can use a fishbone, with the head being the "excess of policies." Key causes will be identified, and data can be collected based on this problem definition.

WHAT WE HAVE LEARNED IN CHAPTER 8

1. Getting valid samples and data is just as important as applying any statistical tool.
2. The people and methods used for taking the samples and data affect the randomness and accuracy of both. Also, the product population is changing as the process changes, sometimes quite radically and often.
3. It is generally not possible to sample all the mini-populations caused by people and process changes.
4. Use the fishbone diagram to identify the key process input variables (KPIVs) that created all these mini-populations. Use the problem definition and close analysis of the fishbone to limit your focus.
5. Generally, the easiest approach to improving a process's output quality is to center the total process or change the process average, rather than reducing the variation. For example, if the diameter of a product is running off-center, it is generally easier to get the average back on-center than to reduce the process variation.
6. If the variation is very high, you may have no choice but to attempt to reduce it. However, not only is improving the total process average easier, but the sampling process and valid data issues are minimized. In both cases, use a fishbone diagram to help identify the most stable process (day, shift, operator, product, and so on) to test.

7. Take a statistically valid sample before and after a change to be confi-dent that the improvement was significant. The formulas for minimum sample size are covered later in the book.

8. Once the change is implemented, validate the effect on the total pro-cess. You can use quality department data for this validation if the means of collecting data stays consistent before and after the change. If you can't trust the quality department data, you will have to take additional samples from the populations before and after the change to validate that the predicted improvement truly happened.

RELATED READING

Roger W. Hoerl and Ronald D. Snee, *Statistical Thinking: Improving Business Performance* (Pacific Grove, CA: Duxbury-Thomson Learning, 2002).

CHAPTER 9

Simplified Gauge Verification

W hat you will learn in this chapter is how to determine gauge error and how to correct this error if it is excessive. When a problem surfaces, one of the first things that must be done is to get good data and measurements related to the problem. The issue of gauge accuracy, repeatability, and reproducibility applies everywhere that data measurements of variables are taken.

Before we use data, we must be satisfied that the data are accurate. One of the most frequent sources of error is the device used to measure the product or process. This device can be as simple as a ruler or as complex as a radiation sensor.

Data error can give us a false sense of security—we believe that the process is in control and that we are making acceptable product—or cause us to make erroneous changes to the process. These errors can be compounded by differences between the gauges used by the supplier and those used by the customer or by variation among gauges within a manufacturing plant.

In Six Sigma projects, the use of this simplified gauge verification tool often gives an insight that allows for big gains with no additional efforts. In the DMAIC process, this tool can be used in the Define, Measure, Analyze, Improve, and Control steps.

MAXIMUM GAUGE ERROR

Ideally, a gauge should not "use up" more than 10 percent of the allowable tolerance. Generally, 30 percent is used as a maximum gauge error. If a gauge has a 30 percent error, then the supplier must keep the product measurements within 70 percent of the tolerance to ensure that the product is within specification.

Some plants discover that as many as half of their gauges will not meet the 30 percent criterion. Even after extensive rework, many gauges cannot pass the simplified gauge verification because the tolerance is too tight for the gauge design.

CASE STUDY: GAUGE ERROR IN MEASURING MOLDS

A complex mold was being sourced from an outside supplier that had been supplying these molds for many years. There had been ongoing complaints about the dimensions of the products made from these molds; in response, the mold tolerances had gradually been tightened over the years.

The issue finally became severe enough that a project was undertaken to understand and rectify the problem. The supplier was insistent that its gauge showed that the molds were well within specifications; it even sent the purchasing plant the measurement data. The purchasing plant had only a crude gauge that gave contradictory readings on the complex mold shape. The molds were finally sent out to a firm specializing in three-dimensional measurements to resolve the issue.

This firm specializing in 3-D measurements was able to demonstrate that most of the molds were out of specification, some of them dramatically. The gauge that the mold supplier was using was not capable of measuring the complex shape to the degree of accuracy required; tightening the tolerances had just made the gauge error worse without improving the molds.

Since most plants do not want to run with reduced in-house tolerances, a problem gauge must be upgraded to "use up" less of the tolerance.

The previous case study is discussed further in the chapter on tolerances.

Checking for Gauge Error

There are many methods of checking for gauge error. The method discussed in this text is simple and emphasizes improving the gauge (when required) rather than retraining the inspectors. The reason this approach is taken is that inspectors change frequently, and it is nearly impossible to get someone to follow a very critical procedure routinely to get an accurate gauge reading. It is far better to have a robust gauge that is easy to use and not likely to be used incorrectly.

Gauge verification must include both *repeatability/reproducibility* and *accuracy*. In simple terms, it must check for variations in readings and for the correctness of the average reading.

Repeatability/Reproducibility

These are measurement concepts involving variations in readings. Repeatability is the consistency of measurements obtained when one person measures the same parts or items multiple times using the same instrument and techniques. Reproducibility is the consistency of the average measurements obtained when two or more people measure the same parts or items using the same measuring technique.

Accuracy

Accuracy is a measurement concept involving the correctness of the average reading. It is the extent to which the average of the measurements taken agrees with a true value.

DEFINITION

SIMPLIFIED GAUGE VERIFICATION INSTRUCTIONS

The first step in doing simplified gauge verification is to get several "master" products near the product specification center. The supplier and the customer must agree on the dimensions of these masters.

Masters can be quantified either by using outside firms that have calibrated specialized measuring devices or by getting mutual agreement between the supplier and the customer (Exhibit 9-1).

The reason the gauge error includes the ± accuracy is that the tolerance, which is the reference, also includes the ± variation from the process center.

Using one of the randomly picked masters, have three inspectors (or operators) measure the master seven times each. Have them take the readings as if they were taking them in normal production (amount of time, method used, and so on). (The quality department should have already done its standard gauge setup independent of this verification.) Calculate the average $\bar{x}$ and standard deviation s of *all* 21 readings. If the standard deviation s in the formula that follows is calculated on a manual calculator, use the "$n - 1$" option if it is available.

Just for reference, the reason the s is multiplied by 5 in the formula is that ±2.5 sigma represents 99 percent of the items on a normal distribution. If the effect of accuracy ($2 \times |\text{master} - \bar{x}|$) is the predominant error (compared with the total gauge error), you should examine the setup procedure for the gauge. Accuracy error is normally easier to reduce than repeatability/reproducibility error.

Exhibit 9-1. Visual representation of simplified gauge verification

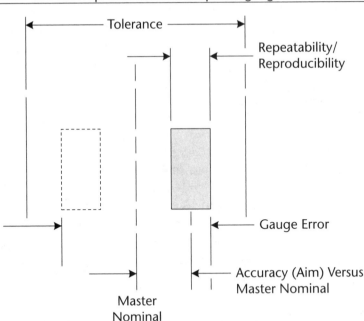

Simplified Gauge Verification, Variables Data

% gauge error = $\dfrac{5s + 2 \times |\text{master} - \bar{x}|}{\text{tolerance}} \times 100$ (ideally <10%, maximum 30%)

FORMULA

$\bar{x}$ = average of *all* 21 readings of a master product
s = standard deviation of *all* 21 readings of the master product
master = standardized dimension of the master product
tolerance = allowable product tolerance (max − min)
$|\text{master} - \bar{x}|$ = difference between master and average $\bar{x}$, ignoring minus sign.

We chose to use a master that was close to the center of the specification, which is sufficient in most cases. However, for some types of gauges, you need to verify readings both at the center and at an end of the specification. Some optical gauges need this verification because mechanical movement of the optics can cause them to read correctly at the center, but not at the end; in this case, masters should be available at a specification end. In most cases, however, masters at the center are sufficient.

Gauge verification must be set up as a regular routine. The simplified gauge verification is in addition to (and independent of) any quality department gauge setup procedure, which is done separately by trained people taking whatever amount of time is required.

TIP

Document Simplified Gauge Verification
Keep documentation of every gauge verification. Make verification and documentation a regular routine.

Example of Simplified Gauge Verification

We want to do the simplified gauge verification procedure on the gauge we are using to measure the diameter on machined shafts. We have established a master shaft, for which the diameter is near nominal. We now have three inspectors (or operators) measure this shaft seven times each. Here are the results.

Master Shaft Standardized Reading = 1.0004 Allowable Shaft Dimensions: max = 1.0050, min = 0.9960, so tolerance = 0.0090 (all dimensions in inches)		
Inspector A	Reading #	Dimension
	1	1.0018
	2	1.0001
	3	0.9998
	4	1.0011
	5	0.9996
	6	1.0001
	7	1.0001
Inspector B	1	0.9992
	2	1.0011
	3	1.0002
	4	0.9991
	5	1.0004
	6	0.9988
	7	1.001
Inspector C	1	0.9997
	2	0.9988
	3	1.0008
	4	1.0015
	5	0.9998
	6	1.0006
	7	0.9996
Average $\bar{x}$ =		1.000152
Standard deviation s =		0.000851
$5s$ =		0.004253

$$\% \text{ gauge error} = \frac{5s + 2 \times |\text{master} - \bar{x}|}{\text{tolerance}} \times 100$$

$$\% \text{ gauge error} = \frac{0.0043 + 2 \times |1.0004 - 1.0002|}{0.0090} \times 100 = 52\%$$

Since the calculated gauge error of 52 percent exceeds the allowable 30 percent, we must go back and find ways to improve the gauge. Assume that we do this. Here are the new results.

Gauge Readings after Rework		
Master Shaft Standardized Reading = 1.0004 Allowable Shaft Dimensions: max = 1.0050, min = 0.9960, so tolerance = 0.0090 (all dimensions in inches)		
Inspector A	Reading #	Dimension
	1	1.0008
	2	1.0006
	3	1.0001
	4	1.0011
	5	1.0004
	6	0.9998
	7	1.0007
Inspector B	1	0.9996
	2	0.9999
	3	1.001
	4	1.0005
	5	1.0001
	6	1.0003
	7	1.0002
Inspector C	1	1.0007
	2	1.0006
	3	0.9996
	4	0.9999
	5	1.0004
	6	1.0006
	7	1.0003
Average $\bar{x}$ =		1.000343
Standard deviation s =		0.000426
$5s$ =		0.002131

$$\% \text{ gauge error} = \frac{5s + 2 \times |\text{master} - \overline{x}|}{\text{tolerance}} \times 100$$

$$\% \text{ gauge error} = \frac{0.0021 + 0.0001}{0.0090} \times 100 = 24.4\%$$

Although the calculated gauge error of 24.4 percent is not great, it is acceptable. Note that by just looking at the readings (before or after rework), it is not obvious whether the gauge is doing an acceptable job, even when we are reading a master with a known diameter. That is why simplified gauge verification often has great payback, because gauge error is a hidden problem (opportunity).

CASE STUDY: GAUGE DIFFERENCES BETWEEN CUSTOMER AND SUPPLIER

A customer was complaining about excursions on a product thickness. To address those concerns, the supplier monitored the next production run very closely, taking many measurements throughout the run. All the measurements showed that the product was well within specifications. The supplier called the customer and told him that he would be pleased with this production run. The supplier even sent the customer the measurements taken during the run.

Once the customer received the product, however, he found that a large number of pieces were clearly out of specification. The customer sent examples of the bad product back to the supplier so that he could see for himself. Again, the supplier measured this returned product as being within specification.

At that point, both the supplier and the customer did simplified gauge verifications, using masters that they agreed upon. The customer's gauge passed, whereas the supplier's gauge did not! The supplier found that his gauge, which was made to measure several similar products, did not readily allow this particular product to seat properly. A new gauge was built for this specific product, and the problem of out-of-specification product was resolved.

It should be noted that when the supplier ran the simplified gauge verification, one of the inspectors *was* able to get the master to read properly. That inspector was aware of the difficulty of seating the product and was able to compensate. However, the supplier very wisely chose to make a new gauge rather than train all the inspectors to seat the product with such precision. Even though it might have been possible to get all the current inspectors to do that, absenteeism and personnel changes made that approach risky as a long-term solution.

Take Advantage of Simplified Gauge Verification
If you have any gauges that take critical measurements and gauge verification has not been done, you have a real opportunity to make a substantial improvement. This is especially true if the customer has reported any problems related to the product or if quality losses are high.

TIP

GAUGE R&R

Most Six Sigma classes teach a process called gauge R&R, but the method taught addresses only repeatability and reproducibility, not accuracy (aim). Gauge R&R uses production products, rather than masters. To get valid results, the current production products must include the full range of process dimensions, which are not always readily available. If the test samples do not include the full range of process dimensions, the gauge R&R often will not pass.

The gauge R&R output is given as an ANOVA (analysis of variance), a rather sophisticated mathematical method that allows gauge error to be separated into operator, gauge, and part contribution. The idea in gauge R&R is that, if the operator is the key contributor to the gauge error, then retraining the person or taking her off the job will correct the problem. If the part is the biggest contributor to the error, it means that something about the physical part, such as distortion, is causing the issue; no direction is given for resolving this issue. If the gauge is the biggest contributor, than the gauge has to be replaced or repaired.

In gauge R&R, accuracy is tested separately, with no effort being made to check the combined effect of repeatability and/or reproducibility and accuracy errors. In fact, gauge R&R doesn't directly address accuracy (aim) at all. It is up to the user of the gauge R&R method to find a way to estimate the total effect of any repeatability/reproducibility error and accuracy error.

The approach taken in this text combines repeatability/reproducibility and accuracy into one test and emphasizes total gauge error. This is based on much experience (frustration) with the difficulty of getting a changing group of inspectors to follow precise instructions for taking measurements. The gauge must be designed for ease of use, so that operator error is minor and is included as part of the total gauge error. Also, the part contribution to gauge error can't be addressed separately in any case, so why separate it?

Anyone who chooses not to use this simplified gauge verification should consider the traditional gauge R&R, since both will offer benefits. If gauge R&R is used, accuracy will have to be checked separately. Obviously, I believe that simplified gauge verification is a better choice.

WHAT WE HAVE LEARNED IN CHAPTER 9

1. In the DMAIC process, simplified gauge verification can be used in the Define, Measure, Analyze, Improve, and Control steps.

2. Ideally a gauge should not "use up" more than 10 percent of the allowable tolerance. The maximum that is acceptable is generally 30 percent.

3. The first step in simplified gauge verification is to get one or more "master" products near the specification center. The supplier and customer must agree on the dimensions of these masters. With some types of gauges (optical), you need to verify readings at both the center and an end of the specification. However, most gauges need masters only at the center of the tolerance.

4. Some production plants discover that as many as half of their gauges will not meet the criterion of maximum 30 percent of tolerance.

5. Emphasize improving the gauge (when required) rather than retraining inspectors. Inspectors come and go, so having them carry out a complex or detailed process is not a good long-term solution.

6. The simplified gauge verification formula includes both repeatability/reproducibility (ability to duplicate a reading) and accuracy (aim or correctness of the average reading).

7. Simplified gauge verification must be set up as a regular routine, and documentation must be kept.

8. Gauge R&R, another approach to verifying that a gauge is not using excessive amounts of the tolerance, is an alternative method. However, it checks for repeatability/reproducibility only. It does not include error caused by inaccuracy, and the user must estimate the combined effect of repeatability/reproducibility error and accuracy error.

RELATED READING AND SOFTWARE

Rath & Strong Management Consultants, *Rath & Strong's Six Sigma Pocket Guide* (Lexington, MA: Rath & Strong/Aon Consulting Worldwide, 2000).

Mark J. Kiemele, Stephen R. Schmidt, and Ronald J. Berdine, *Basic Statistics: Tools for Continuous Improvement*, 4th ed. (Colorado Springs, CO: Air Academy Press, 1997).

MINITAB 13, Minitab Inc., State College, PA; www.minitab.com.

CHAPTER 10

Probability

What you will learn in this chapter of the book is to let the data drive problem solving. However, to interpret data, you need to make a judgment as to whether unusual results were due to a random cause, like someone flipping a fair coin and getting an excessive number of heads by chance, or due to an assignable cause, like the coin having two heads. A knowledge of probability helps you make this determination with a minimum number of samples.

A great deal of Six Sigma work can't be done without some understanding of probability statistics. Probability can be used in all the steps of DMAIC. You will be able to use these techniques to solve many problems in the workplace without using additional tools.

Probability

Manufacturing. On any production line with multiple heads, compare defect levels from each head to see if they are significantly different. Compare two or more similar production lines, shifts, days of the week, and so on. Often you will see significant differences in defect levels that can be addressed at little cost.

Sales. Compare salespeople. The criteria could include new customers, lost sales, or other such factors. Cross-training between the best and worst performers can often improve both! Also, through these careful comparisons, compensation can be made more equitable.

APPLICATIONS

Marketing. Check whether sales increased significantly in multiple markets after a marketing campaign.

Accounting and software development. Compare error incidence to check for significant difference between groups.

Receivables. Check the effect of increased or decreased monitoring of overdue receivables.

Insurance. Compare the complaints at similar-sized treatment centers. The criteria could include patient care, billing errors, or other such factors.

CASE STUDY: APPLICATION OF SIMPLE PROBABILITY

A production plant was in trouble. It was getting multiple defects in the product going out from every production line. The managers felt that the issue had to be caused by a problem with the incoming raw material. They had given up on making good product and made a panicked call for help to their home-office engineering group. A task force was assembled and began attacking the problems.

One of the first things an engineer did was collect defects from one line. He quickly concluded that one of the defects was coming from one set of tooling. (There were 20 tooling sets per line.) He had the tooling set taken out of production and found that it had been assembled wrong. Upon looking at maintenance records for the machine, he found that the tooling set had been on the machine for two weeks. Since there were 20 tooling sets on the line, 5 percent of the product coming from the machine for two weeks had been defective, since every product coming from this set of tooling was bad.

The engineer knew that raw materials would not cause this 100 percent defect rate on only one set of tooling without also affecting products from the other tooling sets. This was what triggered him to have the tooling set removed and inspected. Using this kind of systematic analysis of defect data and reacting accordingly, the plant was back to normal productivity within three days. There had been no problem with the raw materials.

Certainly this reasoning took only a rudimentary knowledge of probability, but no one in the plant, including the plant engineers, was analyzing the problem in this way. They just assumed that, since they were having multiple problems throughout the plant, the issue must be raw materials.

At the end of the third day, the task force received a call from its general manager, who wanted an update on how the team was doing. When the team leader told the GM that the plant was now back to targeted production using the very basic analysis just described, the GM initially did not believe it! It seemed too simple. The task force went home.

By the beginning of the following week, the plant yields had again slipped dramatically. This time, only one of the task force engineers returned and, using the same methods, was again able to get the plant back to targeted production within three days.

The net effect of all this was that the plant manager at the plant was removed from his job, and the plant began to pay more attention to the required detail of running production.

The case study just given is typical in that the initial conclusion that the problem was caused by the raw materials was made without carefully analyzing the data. In contrast, the engineer used specific data and an elementary knowledge of probability to reach his conclusions, since he knew that defects caused by raw materials would have been random, not specific to one of the sets of tools on a line. Although the other production lines did not have the identical problems, the same kind of careful analysis based on detailed data resolved the production problems.

USES OF PROBABILITY

This chapter on probability will help you solve problems when the data aren't quite as obvious. Not only is the probability analysis included the basis for many of the Six Sigma tools, but knowledge of probability can also be used independently on many problems.

We frequently make estimates concerning the likelihood of an event, or its probability—for example, the chance of rain, the chance of winning the lottery, or the chance of being in a plane crash. Some probabilities are easy to calculate and intuitive, like the chance of getting a head on a coin flip (one in two, or 0.5). Some probabilities are neither easy to calculate nor intuitive, like the probability of an earthquake.

In Six Sigma, we also need to estimate the probability of an event. In this way, we can make some judgment as to whether something happened because of a random coincidence or whether there is an assignable cause that we should address. Luckily, the work you will be doing does *not* involve earthquakes!

We will start with problems where we know the mathematical probability of a single random event, verifying any answer with both the abbreviated binomial table (Exhibit 10-1) and Excel's BINOM.DIST (binomial distribution) statistics option. Excel's BINOM.DIST is the primary tool you will use to solve these problems, but if you use the abbreviated binomial table first, you will get a more fundamental understanding of probability. This understanding will minimize the likelihood of an error when using any statistics software package.

Probability (BINOM.DIST)

n = the number of independent trials, like the number of coin tosses, the number of parts measured, and so on.

Probability p (or probability s) = the probability of a "success" on *each individual trial*, like the likelihood of a head on one coin flip or a defect on one part. This is always a proportion and is generally shown as a decimal, like 0.0156.

Number s (or x successes) = the total number of "successes" that you are looking for, like getting exactly three heads.

Probability P = the probability of getting a given number of successes from *all the trials*, like the probability of three heads in five coin tosses or 14 defects in a shipment of parts. This is often the answer to the problem.

Cumulative = the sum of the probabilities of getting "the number of successes or fewer," like getting three *or fewer* heads on five flips of a coin. This option is used on "less than" and "more than" problems.

These definitions and their uses will become apparent as you solve the following problems.

Problem 1
What are the chances of getting three heads in three flips of a coin?

Using the previous definitions, recognize that n (number of flips) = 3. The probability of getting a head on any one flip is $p = 0.5$. The number of "successes" is 3 (3 heads). P is the probability of getting exactly 3 heads on 3 coin flips, which is the desired answer to the problem.

We can make a table of all possible equally likely outcomes of the three coin flips:

Outcome	Flip 1	Flip 2	Flip 3
1.	head	head	head
2.	head	head	tail
3.	head	tail	head
4.	head	tail	tail
5.	tail	head	head
6.	tail	head	tail
7.	tail	tail	head
8.	tail	tail	tail

As you can see, only one of the eight equally likely outcomes of 3 coin flips is 3 heads. So, $P = 1/8$, or 0.125.

You can get the same answer logically (or mathematically). The chance of getting heads on the first flip is 50 percent, or $p = 0.5$. On each successive flip, the chances of a head are also 0.5. So, the chance of getting 2 heads on 2 coin flips is $0.5 \times 0.5 = 0.25$. Similarly, since each trial is independent (not affected by earlier tosses), we can multiply the probabilities on 3 flips: $0.5 \times 0.5 \times 0.5 = 0.125$.

Verify this answer by using the abbreviated binomial table in Exhibit 10-1. Note that more complete tables are available in most statistics books.

Exhibit 10-1. Abbreviated binomial table. Values within the table are the probability of getting exactly x successes on n trials.

n $\#$ of trials	x successes on n trials	p (each trial) =	0.125 (1/8)	0.167 (1/6)	0.250 (1/4)	0.500 (1/2)
2	0		0.7656	0.6944	0.5625	0.2500
2	1		0.2188	0.2778	0.3750	0.5000
2	2		0.0156	0.0278	0.0625	0.2500
		Sum of P:	1.0000	1.0000	1.0000	1.0000
3	0		0.6699	0.5787	0.4219	0.1250
3	1		0.2871	0.3472	0.4219	0.3750
3	2		0.0410	0.0694	0.1406	0.3750
3	3		0.0020	0.0046	0.0156	0.1250
		Sum of P:	1.0000	1.0000	1.0000	1.0000
4	0		0.5862	0.4823	0.3164	0.0625
4	1		0.3350	0.3858	0.4219	0.2500
4	2		0.0718	0.1157	0.2109	0.3750
4	3		0.0068	0.0154	0.0469	0.2500
4	4		0.0002	0.0008	0.0039	0.0625
		Sum of P:	1.0000	1.0000	1.0000	1.0000
5	0		0.5129	0.4019	0.2373	0.0313
5	1		0.3664	0.4019	0.3955	0.1563
5	2		0.1047	0.1608	0.2637	0.3125
5	3		0.0150	0.0322	0.0879	0.3125
5	4		0.0011	0.0032	0.0146	0.1563
5	5		0.0000	0.0001	0.0010	0.0313
		Sum of P:	1.0000	1.0000	1.0000	1.0000
10	0		0.2631	0.1615	0.0563	0.0010
10	1		0.3758	0.3230	0.1877	0.0098
10	2		0.2416	0.2907	0.2816	0.0439
10	3		0.0920	0.1550	0.2503	0.1172
10	4		0.0230	0.0543	0.1460	0.2051
10	5		0.0039	0.0130	0.0584	0.2461
10	6		0.0005	0.0022	0.0162	0.2051
10	7		0.0000	0.0002	0.0031	0.1172
10	8		0.0000	0.0000	0.0004	0.0439
10	9		0.0000	0.0000	0.0000	0.0098
10	10		0.0000	0.0000	0.0000	0.0010
		Sum of P:	1.0000	1.0000	1.0000	1.0000

Here's how to use the Abbreviated Binomial Table (Exhibit 10-1) on problem 1.

Find $n = 3$ in the leftmost column, which is the number of trials.

Find the number of successes (3 heads).

In the far column, $p = 0.500$ (chance of a head on each flip), the value 0.1250 is P.

Now do it in Excel. After bringing up the Excel worksheet, click on "Formulas," "More functions," "Statistical," then "BINOM.DIST."

In the first box, enter the "number of successes" (number of heads) you want in these trials, which is "3." The second box asks for the "number of trials," which is "3." The third box asks for the probability of a "success" (head) on each trial, which is "0.5."

The fourth box asks if the problem requires the cumulative probability. If you answer "true," you get the sum of the probabilities up to 3 (the probability of 0 heads + the probability of 1 head + the probability of 2 heads + the probability of 3 heads), which is the probability of getting 3 or fewer heads. In this problem, you do *not* want the cumulative probability, so answer "false." You then get our desired probability of exactly 3 heads, which is $P = 0.125$.

So, here is a summary of what we just did:

Excel BINOM.DIST
successes = 3
trials = 3
probability $(p) = 0.5$
cumulative: false
The result is $P = 0.125$.

Problem 2
What is the probability of getting 2 or fewer heads in 3 flips of a coin?

We will show five ways to get the answer to this problem.

You could look at the possible outcomes table (page 81) and add up the number of outcomes with 0 heads (1 outcome), 1 head (3 outcomes), and 2 heads (3 outcomes), for a total of 7 outcomes (out of 8 possible outcomes). This is a probability of 7/8, or 0.875.

The same answer can be found using the abbreviated binomial table (Exhibit 10-1) by adding the probabilities of 0 successes (0.125), 1 success (0.375), and 2 successes (0.375), for a total = 0.875.

Equally, using the BINOM.DIST function in Excel, you could do it three times, adding the results of 0, 1, and 2 successes (with cumulative: false), and you get $0.125 + 0.375 + 0.375 = 0.875$.

Or you can recognize that the only outcome that has more than 2 heads is 3 heads, which has a probability of 0.125 (from Problem 1). Since the sum of the probabilities of all possible outcomes always equals 1 (see the abbreviated binomial table, Exhibit 10-1), we can subtract the probability of 3 heads (0.125) from 1 to get the answer = 0.875!

The following is the most direct way to the answer. Using Excel BINOM.DIST, we can realize that the "cumulative true" gives the probability of getting 2 or fewer heads, which is what we want. So, here's how we can get the answer directly:

Excel BINOM.DIST
successes = 2
trials = 3
probability (p) = 0.5
cumulative: true
The result is $P = 0.875$.

Use the Sum of Probabilities = 1

TIP

Since the sum of the probabilities of all possible outcomes always equals 1, we can often use this knowledge to simplify a problem.

For example, if we want to know the probability of getting 1 or more heads on 10 coin tosses, we can find the probability of getting 0 heads, then subtract this probability from 1. This is much easier than adding the probabilities of getting 1 head + 2 heads + 3 heads + 4 heads + 5 heads + 6 heads + 7 heads + 8 heads + 9 heads + 10 heads.

Using Excel's BINOM.DIST Cumulative Function

TIP

Do not use the cumulative function for the probability of a single outcome/ success, like 3 heads out of 5 coin tosses. Enter "false" in the box for cumulative function.

Use the cumulative function for "less than," "equal to or less than," "equal to or greater than," or "greater than" a given outcome or success, as follows.

For "less than" a given outcome, like fewer than 3 heads out of 8 coin tosses, use the cumulative function "true" with the successes at one less than the given value (3 – 1 = 2).

successes = 2
trials = 8
p = 0.5
cumulative: true
The result is $P = 0.1445$.

For "equal to or less than" a given outcome, like 3 heads or fewer out of 8 coin tosses, use the cumulative function "true" with the successes at the given value (3).

successes = 3
trials = 8
$p = 0.5$
cumulative: true
The result is $P = 0.3633$.

For "greater than" a given outcome, like more than 3 heads out of 8 coin tosses, use the cumulative function "true" with the successes at the given value (3), then subtract the result from 1.

successes = 3
trials = 8
$p = 0.5$
cumulative: true
The result is 0.3633. P would then equal $1.0000 - 0.3633 = 0.6367$.

For "equal to or greater than" a given outcome, like 3 or more heads out of 8 coin tosses, use the cumulative function "true" with the successes at one less than the given outcome ($3 - 1 = 2$), then subtract the result from 1.

successes = 2
trials = 8
$p = 0.5$
cumulative: true
The result is 0.1445. P would then equal $1.0000 - 0.1445 = 0.8555$.

You can satisfy yourself that these cumulative function examples on 8 coin tosses make sense by seeing that the probability of "less than 3 heads" plus the probability of "3 or more heads" equals 1. The same is true for "3 heads or fewer" plus "more than 3 heads."

Problem 3
What is the chance of getting 8 or fewer tails on 10 flips of a coin?

Using the abbreviated binomial table (Exhibit 10-1), we can add the probabilities of getting 0, 1, 2, 3, 4, 5, 6, 7, and 8 tails. However, it is easier to add the probabilities of getting 9 and 10 tails, then subtract from 1. The probability of 9 tails is 0.0098, and the probability of 10 tails is 0.0010. Adding these and then subtracting the sum from 1 gives $1.0000 - 0.0108 = 0.9892$. So, the P of getting 8 or fewer tails on 10 coin flips is 0.9892, or 98.92 percent.

Or, here's the most direct way:

Excel BINOM.DIST
successes = 8
trials = 10
$p = 0.5$
cumulative: true
The result is $P = 0.9893$ (the slight difference from the previous result is due to rounding error).

CASE STUDY: EXCESSIVE ACCIDENT RATE

A sales force had an accident rate that was excessive, and the sales manager was under a lot of pressure to reduce it. One of the salespeople thought that the accident rates were higher near the end of the year, so he looked at 100 random accidents from each of several years and calculated the average accident rate for each month. He found that December had the highest average of any month, at 15 accidents. The other months all had lower rates, and the rates in those other months were more or less similar to each other.

The sales manager determined that the chance of getting 15 or more accidents in December because of random causes alone was very unlikely. Upon analyzing further, he also noted that the accident rate for January was not high, so he doubted that weather was the cause.

The next year, the sales manager dictated that the salespeople give gifts to the customers during the holiday season, rather than partying with them, and the accident rate went down to the same level as the other months.

Let's verify the sales manager's finding that December's 15 accidents out of 100 for a year was "very unlikely." First, some interpretation of *what* was "very unlikely." The sales manager wasn't surprised that the December accident average was *exactly* 15; he was surprised that it was that high. In fact, 16 or 17 accidents would also have surprised him! So, that is why he checked against the likelihood of 15 *or more* accidents happening in December as a result of random causes. This approach is more conservative and made him less likely to erroneously find that something happened because of other than random causes. The likelihood of getting an *exact* number is generally small and will bias results to show that the result was not random.

Excel BINOM.DIST

successes = 14

trials = 100

$p = 0.08333$ (1/12, which is what would be expected for a random month)

cumulative: true

The result is 0.9814. P would then equal $1.0000 - 0.9814 = 0.0186$, or 1.86 percent.

So the sales manager was correct in saying that 15 or more accidents in December was very unlikely, since it would be expected to happen randomly only 1.86 percent of the time. He would therefore be 98.14 percent (100 percent − 1.86 percent) confident that the December accident rate was not random.

Note that it was not sufficient just to find that the December accident rate was higher than the $100/12 = 8.33$ he would have expected for a random month. It had to be determined with some confidence that the December results were high enough that they were probably not random.

The sales manager did not try to determine the root cause of December's higher accident rate. It could have been because of excessive drinking, more miles driving visiting customers, driving more at night, or other reasons. He saw only the correlation between the month and the accident rate. His chance of success in implementing a solution would have been higher if he could have identified the root cause.

Independent Trials Are Not Affected by Earlier Results

TIP

The probability for an independent trial is *not* affected by results on earlier trials. For example, someone could flip 10 heads in a row, but the probability of a head on the next coin flip is still $p = 0.5$ (assuming that both the coin and the person tossing it are honest and so on).

Problem 4

What are the chances of getting at least 8 tails on 10 coin flips?

We could use either the abbreviated binomial table (Exhibit 10-1) or Excel BINOM.DIST to find the P of 8, 9, and 10 tails. We would then add these to get the total probability $P = 0.0439 + 0.0098 + 0.0010 = 0.0547$.

Or, using the Excel BINOM.DIST and realizing that the chance of at least 8 tails (or 8 or more tails) is the same as 1 minus the chance of 7 or fewer tails, we solve as follows:

Excel BINOM.DIST

successes = 7

trials = 10
$p = 0.5$
cumulative: true
The result is 0.9453. P would then equal $1.0000 - 0.9453 = 0.0547$, or 5.47 percent.

Problem 5

A vendor is making 25 percent defective product. In a box of 10 random parts from this vendor, what is the probability of finding 2 or fewer defects?

If we use the abbreviated binomial table (Exhibit 10-1), with $n = 10$, successes = 2, 1, and 0, and $p = 0.25$, we get 0.282, 0.188, and 0.056. We add these and get $P = 0.526$, or 52.6 percent.

Or, using Excel BINOM.DIST:

successes = 2
trials = 10
$p = 0.25$
cumulative: true
The result is $P = 0.526$, or 52.6 percent.

Problem 6

A salesperson has been losing 25 percent of potential sales. In a study of 10 random sales contacts from this salesperson, what is the probability of finding 3 or more successful sales?

We must be careful that what we call a "success" (*successful* sales in this case) is consistent with the rest of the problem statement (which is currently stated as percent of *lost* sales). We can restate the question as "getting 75 percent of potential sales" so that the success is measured in the same terms as the rest of the problem statement. Then we have the following problem:

A salesperson has been successful in getting 75 percent of potential sales. In a study of 10 random sales contacts from this salesperson, what is the probability of finding 3 or more successful sales?

Again, to save work, we know that "3 or more successful sales" is the same as 1 minus the probability of "2 or fewer successful sales."

Excel BINOM.DIST
successes = 2
trials = 10
$p = 0.75$

cumulative: true
The result is 0.0004. P would then equal 1.0000 − 0.0004 = 0.9996, or 99.96 percent.

Or, we can restate the problem in terms of "lost sales":

A salesperson has been losing 25 percent of potential sales. In a study of 10 random sales contacts from this salesperson, what is the probability of finding 7 or fewer lost sales?

Note that in this problem restatement, we had to recognize that 3 or more successful sales is the same as 7 or fewer lost sales. If this is not obvious, consider that when you have 3 successes out of 10 potential sales, 7 are lost; with 4 successes, there are 6 losses; with 5 successes, there are 5 losses; with 6 successes, there are 4 losses; and so on.

Excel BINOM.DIST
successes = 7
trials = 10
$p = 0.25$
cumulative: true
The result is $P = 0.9996$, or 99.96 percent, which is consistent with the previous answer.

This problem shows that if you think your problem through carefully and are careful that the definition you've chosen for "success" is consistent with the other values used in the solution, you will get a correct and consistent answer.

CASE STUDY: DEFECTS PRIMARILY IN ONE QUADRANT

A high-speed production line was making a small number of critical defects. A quality engineer randomly collected 100 defects. He examined them and found that 22 defects were from the first quadrant of the product, 36 from the second, 21 from the third, and 21 from the fourth. He concluded that there was less than a 1 percent chance that 36 or more defects out of 100 would be coming from the second quadrant because of random causes alone.

So he went looking for anything suspicious on the production line that was limited to the second quadrant. He found a cooling nozzle from which some of the spray hit the second quadrant of the product. Since the product was very hot at that point in the process, he suspected that stress was being introduced.

When he asked the operator why he had put the spray at that location, he was told that it was being used to cool the tooling that was adjacent to the product at this point. The engineer then designed a more directed spray method that missed the product but hit the tooling, which allowed adequate cooling and also solved the problem of excess defects in that quadrant.

Let's see if we agree with the engineer's conclusion that the higher number of defects in the second quadrant was probably not due to random causes alone. ("Random" would mean that there was nothing peculiar about this quadrant: he just happened to get a sample with more defects in this location.) Note that he checked the chances of 36 *or more* defects happening randomly, since there was nothing special about there being *exactly* 36. So, we want to calculate the chances of 36 or more defects occurring in the second quadrant from random causes only.

Excel BINOM.DIST
successes = 35
trials = 100
$p = 0.25$ (1/4, the probability of the defect randomly occurring in the second quadrant)
cumulative: true
The result is 0.9906. *P* would then equal $1.0000 - 0.9906 = 0.0094$, or 0.94 percent.

So, there is only a 0.94 percent chance of getting 36 or more of the 100 defects in the second quadrant because of random causes. Thus, the engineer was right to be suspicious—and his conclusion enabled him to focus his attention on areas affecting only that quadrant, which minimized the areas in the process that he had to examine to identify the source of the problem.

Excel BINOM.DIST Trials Max at 1,000

TIP

Excel BINOM.DIST allows a maximum of 1,000 trials. So, if you have more than 1,000 trials, proportion the trials and the number of successes to 1,000.

For example, if the data have 2,000 trials and you are looking for 140 successes, use Excel BINOM.DIST with 1,000 trials and 70 successes. The *p* is not affected.

Problem 7
These are the results of 109 random erroneous orders processed by seven telephone operators, A through G:

Operator	Erroneous Orders
A	14
B	16
C	22
D	13
E	16
F	15
G	13

Is Operator C's performance worse than we should expect, since C's error total of 22 is higher than that of any of the other six operators? We would want to be at least 95 percent confident that the performance was poor before we took any action. We want to check against the likelihood of getting 22 *or more* errors, rather than exactly 22. This will be more conservative, and it's not important that there are *exactly* 22 order errors.

Excel BINOM.DIST
successes = 21
trials = 109
$p = 1/7 = 0.14286$ (random chance, since there are seven operators)
cumulative: true
The result is 0.9429. *P* would then equal 1.0000 − 0.9429 = 0.0571, or 5.7 percent.

This means that there is a 5.7 percent chance of this happening randomly without an assignable cause. The confidence level of the conclusion that this is not random would therefore be 100 percent − 5.7 percent = 94.3 percent, which is below the 95 percent test threshold.

Therefore, we can't conclude with 95 percent confidence that Operator C is performing worse than any of the other operators.

Additional Practice Problems

Problem 8
What is the likelihood of getting 4 heads in 7 flips of a coin?

Problem 9
What is the likelihood of getting at least 4 heads in 7 flips of a coin?

Problem 10

What is the probability of getting 3 fives on 6 rolls of a die?

Problem 11

What is the probability of getting 3 or more fives rolling 6 dice one time?

Problem 12

An automaker entered the marketplace with a car to compete with two other brands that were already in that market. In the first month after its introduction, the new entry got 355 sales out of a random sample of 1,000 sales from the total market. Can the automaker with the new entry say with 95 percent confidence that he got more than the projected 1/3 of the sales due to other than random causes?

Problem 13

An automatic transmission repair shop had an established standard for doing a certain type of repair. The standard was based on the historic average time mechanics had taken to do a similar repair. The shop hired a new mechanic. After six months, they sampled 12 of the new mechanic's repair times and compared them to the standard repair times. The new mechanic took more than the standard time on 8 of the 12 samples. How confident would the repair shop be in judging that the higher times were due to performance rather than to random causes?

Solutions to Additional Practice Problems

Problem 8

What is the likelihood of getting 4 heads in 7 flips of a coin?

Excel BINOM.DIST
successes = 4
trials = 7
$p = 0.5$
cumulative: false
The result is $P = 0.2734$, or 27.34 percent.

Problem 9

What is the likelihood of getting at least 4 heads in 7 flips of a coin?

Excel BINOM.DIST
successes = 3
trials = 7
$p = 0.5$
cumulative: true
The result is 0.500. P would then equal $1.000 - 0.500 = 0.500$, or 50.0 percent.

Problem 10
What is the probability of getting 3 fives on 6 rolls of a die?

Excel BINOM.DIST
successes = 3
trials = 6
$p = 1/6 = 0.1667$
cumulative: false
The result is $P = 0.0536$, or 5.36 percent.

Problem 11
What is the probability of getting 3 or more fives rolling 6 dice one time?

Excel BINOM.DIST
successes = 2
trials = 6
$p = 1/6 = 0.1667$
cumulative: true
The result is 0.9377. P would then equal $1.0000 - 0.9377 = 0.0623$, or 6.23 percent.

   (Note that rolling 6 dice at one time is the same as rolling 1 die 6 times. This is because in both cases, the result on each die is independent of the results on the others.)

Problem 12
An automaker entered the marketplace with a car to compete with two other brands that were already in that market. In the first month after its introduction, the new entry got 355 sales out of a random sample of 1,000 sales from the total market. Can the automaker with the new entry say with 95 percent confidence that he got more than the projected 1/3 of the sales due to other than random causes?

   We want to check against the likelihood of 355 *or more* sales, since getting *exactly* 355 is not what is important!

Excel BINOM.DIST
successes = 354
trials =1,000
$p = 1/3 = 0.3333$
cumulative: true
The result is 0.9220. P would then equal $1.0000 - 0.9220 = 0.0780$, or 7.8 percent.

The automaker's confidence would therefore be 100.00 percent – 7.8 percent = 92.20 percent, which is below the 95 percent confidence target level the automaker wanted. Therefore, the automaker *can't* claim that the new entry gained more than 1/3 of the market because of other than random cause.

Note that if this problem had been done using *exactly* 355 as the test criterion, it would have shown that the automaker *did* get more than 1/3 of the market with 95 percent confidence. This type of error occurs because getting *any* specific result randomly with this number of trials is unlikely.

Problem 13

An automatic transmission repair shop had an established standard for doing a certain type of repair. The standard was based on the historic average time mechanics had taken to do a similar repair. The shop hired a new mechanic. After six months, they sampled 12 of the new mechanic's repair times and compared them to the standard repair times. The new mechanic took more than the standard time on 8 of the 12 samples. How confident would the repair shop be in judging that the higher times were due to performance rather than to random causes?

Assume that an average mechanic's repair times would be above standard 50 percent of the time, so $p = 0.5$. As in the previous problem, we want to check versus 8 *or more* repair times over the standard, since having *exactly* 8 is not the issue.

Excel BINOM.DIST
successes = 7
trials = 12
$p = 1/2 = 0.5$
cumulative: true
The result is 0.8062. P would then equal $1.0000 - 0.8062 = 0.1938$, or 19.38 percent.

So, this result would happen randomly 19.38 percent of the time. Therefore, the repair shop would be only 100 percent – 19.4 percent = 80.6 percent confident that the new mechanic was taking more time than the standard because of other than random causes.

WHAT WE HAVE LEARNED IN CHAPTER 10

1. Six Sigma work can't be done without some understanding of probability statistics.
2. Letting data drive your problem solving will often save work and lead to a condensed list of solution considerations.
3. Using basic probability techniques on the data will help separate random results from assignable causes.
4. Since the sum of the probabilities of all possible outcomes always equals 1, we can often use this knowledge to simplify a problem.
5. Excel BINOM.DIST is sufficient for doing the probability calculations, using the cumulative function to save work in doing less than or greater than calculations.
6. Examine the data carefully to make sure they are truly independent.
7. Although probability techniques help to focus on solution options, additional analysis or trials are usually needed to identify a specific cause-and-effect relationship.
8. Because our confidence level is *never* 100 percent, any change must be verified with future tests or data. Past data can never validate a change.
9. You can already do real Six Sigma work by applying these basic probability techniques. Even if you read no further in this book, just applying this chapter will enable you to make substantial and meaningful improvements in diverse applications.

RELATED READING

Mark J. Kiemele, Stephen R. Schmidt, and Ronald J. Berdine, *Basic Statistics: Tools for Continuous Improvement*, 4th ed. (Colorado Springs, CO: Air Academy Press, 1997).

Marek Capinski and Tomasz Zastawniak, *Probability Through Problems* (New York: Springer, 2001).

David M. Levine, David Stephan, Timothy C. Krehbiel, and Mark L. Berenson, *Statistics for Managers, Using Microsoft Excel* (with CD-ROM), 4th ed. (Upper Saddle River, NJ: Prentice Hall, 2004).

CHAPTER 11

Data Plots
and Distributions

W hat you will learn in this chapter of the book is how to plot data and how to spot opportunities from these plots. You will also learn what a *normal* distribution of data is and some terminology that describes this distribution. Like that in the previous section, this information will help you to solve many real problems and is needed for Six Sigma work. Plotting data is a necessary step in implementing many of the Six Sigma tools. It is used in all the steps of the DMAIC methodology.

Receivables. Plot the monthly receivables year over year and compare the data with cash flow.

Insurance. Compare surgeries done in comparable hospitals to spot cost differences.

CASE STUDY: COMPARING PLOTS OF TWO PRODUCTION LINES

A production plant had two similar lines producing containers. The wall thickness on these containers was critical to the customer, so measurements were taken regularly and entered into a computer file.

The customer had periodically expressed a preference for containers from line 2 over containers from line 1, but the customer had no data to substantiate this preference. Since both lines made product that was within specifications, the container plant felt that any difference was imagined, as the two lines were thought to be identical. Finally, an engineer plotted 1,000 random wall thickness measurements from each line.

Exhibit 11-1 shows the histogram plots of the data from the two lines, with one histogram overlaid on the other for ease of comparison.

Exhibit 11-1. Histogram of two production lines

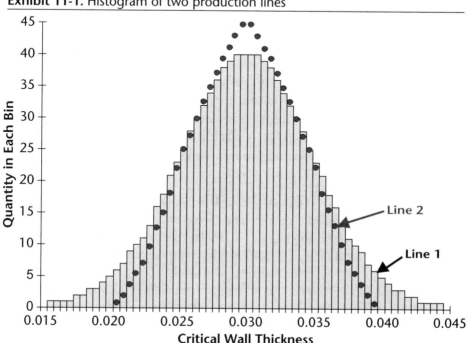

As you can see by looking at the histograms in Exhibit 11-1, the wall thickness measurements from line 1 are more dispersed (more high and low values) than those from line 2. This is why the customer was happier with the containers from line 2.

Using these data as a motivator, the engineer was able to find subtle differences between the two lines and then eliminate those differences. The wall thickness of containers from line 1 became nearly identical to the wall thickness of containers from line 2. The customer saw $25,000 per year savings from the resultant improved product, and continuing plots of the data after the changes substantiated that the two lines were now making nearly identical product.

In this case study, there are several things to be noted. First, the customer's feelings on quality were ignored because there were no supportive data and because both lines were making product that was within specifications. Second, the data were already available in a database that no one had bothered to examine. Third, although both lines made product that was within specifications, the customer saw the improvement in the revised process. Fourth, without anyone realizing it, the two lines were not identical, and over time small changes had been incorporated and had not been documented. Once someone decided to plot the data, it was obvious that the customer was correct and that the products from the two lines were not the same.

NORMAL DATA

A lathe is machining shafts to a 1.0000″ nominal diameter. You carefully measure the diameters of 100 of these shafts. If you sort the diameters into 0.0005″-wide "bins" and plot these data, you will get a *histogram* similar to that shown in Exhibit 11-2. In this illustration, the ends of the shafts are shown for clarity. This would not be true in a regular histogram.

In a Histogram, Assume No Values Are on Bin "Edges"   TIP
When a value appears to be exactly on a bin "edge," the convention is to put that value into the higher bin. In the following example, if a shaft were measured to be *exactly* 1.0000″, it would be put into the 1.0000″-to-1.0005″ bin.

Not having a value on the bin edge is not difficult to accept when you consider that with an accurate enough measurement system, you would be able to see even the smallest difference from exactly 1.0000″.

Exhibit 11-2. Histogram of 100 shafts

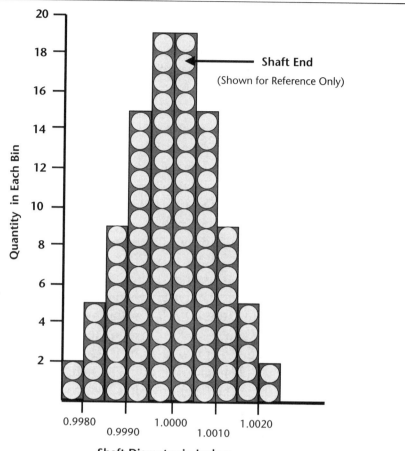

Shaft End
(Shown for Reference Only)

Quantity in Each Bin

Shaft Diameter in Inches
(100) 1" Nominal Shafts Sorted into 0.0005" Bins

We must now interpret the data shown in this histogram.

First, notice that the process is centered and that the left half is a mirror image of the right. What percent of the shafts are within 0.001″ of the 1.000″ nominal diameter? Adding the bin quantities on both sides of the center ±0.001″, we get 68, or 68 percent of the shafts. It will be shown later that on any process with a normal distribution, this 0.68 point (or 0.34 on either side of the center) is equal to ±1 sigma (or ±1 standard deviation).

For illustration purposes, 1 sigma in this case just happens to equal 0.001″. Therefore, 2 sigma = 0.002″. What percent of the shafts are within ±2 sigma of the nominal diameter? Again, counting the bin quantities within ±0.002″ on both sides of the center, we find that 96 shafts, or 96 percent of the shafts, are within 2 sigma of the nominal diameter.

Reference Data Within a Normal Distribution   TIP

It is handy to remember that approximately 2/3 (68 percent) of the data points are within ±1 sigma of the center in a process with a normal distribution, and that 95 percent are within ±2 sigma. Another good reference number is that 99.7 percent of the data points are within ±3 sigma of the center.

All of the previous questions referred to data on both sides of the center. However, it is often important to know what is occurring on only one end of the data. For example, what percent of the shafts are at least 1 sigma greater than 1.0000" in diameter? Adding the bin quantities to the right of +1 sigma (sigma in this case happens to be +0.0010"), we get 16, or 16 percent.

We will be using charts (and computer programs) that take the reference points either at the center or at either end of the data. You have to look carefully at the data and the chart illustration to see what reference point is being used.

Now, using some of the techniques from the previous chapter on probability and assuming independence (assume that you put back the first shaft before you pick the second), what is the likelihood of randomly picking two shafts that are above 1.0000" in diameter? Since the probability of each is 0.5, the probability of two in a row is $0.5 \times 0.5 = 0.25$.

The previous example used shafts, but other items could have been plotted with similar results. The height of adult men could have been plotted, with the bins representing 1" height increments. Multiple sales results could be shown as a histogram, with each bin increasing $10,000. Clerical errors could be displayed, with each bin being an increment of errors per 10,000 entries. Stock fund performance could be shown, with the bins being percent annual gain. In all these cases, you will probably get a normal distribution.

Let's now plot the same population of shaft data using 1,000 shafts and breaking the data into 0.0001"-wide bins (Exhibit 11-3).

As we get more data from this process and use smaller bins, the shape of the histogram approaches a *normal* distribution. In fact, it helps to think of a normal "curve" as a normal distribution with very small bins. This is the shape that will occur on many processes.

Exhibit 11-4 gives a normal distribution curve showing how it varies with different values of sigma.

As stated previously, on any plot of data from a normal process, approximately 2/3 (68 percent) of the data points are within ±1 sigma on either side of the center, 95 percent are within ±2 sigma on either side, and 99.7 percent are within ±3 sigma of the center. The use of normal curve standardized data allows us to make predictions on processes with normal

Exhibit 11-3. Histogram of 1,000 shafts

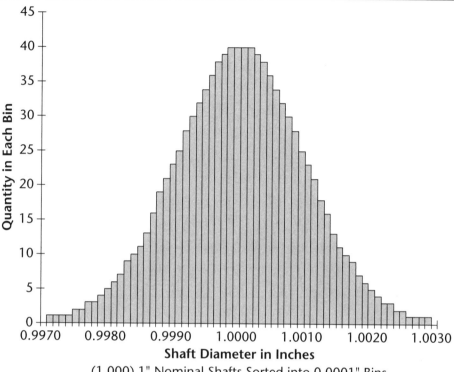

Shaft Diameter in Inches
(1,000) 1" Nominal Shafts Sorted into 0.0001" Bins

Specifying a Normal Distribution

All that is needed to define a normally distributed set of data is the mean (average) and the standard deviation (sigma).

We could calculate the standard deviation (sigma) values manually, but since most $10 calculators and many computer programs calculate these values so easily, we will not calculate them manually. If you use a calculator to do this calculation, you may have your choice of using n or $(n-1)$. Use $(n-1)$.

Just for reference, here's the formula to solve for the standard deviation s on a set of n values, where $\bar{x}$ is the average of all the data points x:

$$s = \sqrt{\frac{(x_1-\bar{x})^2 + (x_2-\bar{x})^2 + \ldots + (x_n-\bar{x})^2}{n-1}}$$

The standard deviation is a measure of the spread of the normal curve. The greater the sigma, the more distributed the data, with more highs and lows.

Exhibit 11-4. Normal distribution with various sigma values

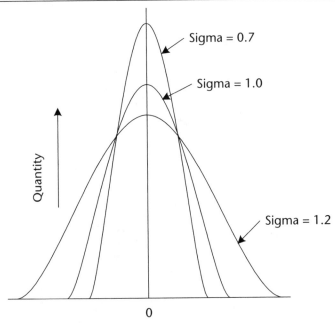

distributions using small samples rather than collecting hundreds of data points on each process. As you will see later, once we establish that a process has a normal distribution, we can assume that this distribution will stay normal unless a major process change occurs.

We will be doing a lot of analysis based on the likelihood of randomly finding data beyond ±2 sigma, or outside of the expected 95 percent of the data. In the case of our 1,000 shafts, Exhibit 11-5 shows our histogram with this 5 percent area darkly shaded on the two ends (below 0.9980″ and above 1.0020″).

To get a sense of what this kind of distribution would look like if the outlying 5 percent were distributed randomly, Exhibit 11-6 shows several hundred shafts with 5 percent of the shafts shaded.

If you picked a shaft randomly from the distribution in Exhibit 11-6, you would be unlikely to pick a shaded one. In fact, if you picked a shaded shaft very often, you would probably begin to wonder whether the distribution really had only 5 percent shaded shafts. Much of the analysis we will be doing uses similar logic.

Let's pursue this further. Suppose you had been led to believe that a distribution had 5 percent shaded shafts, but you suspected that this was not true. If you picked one shaft randomly and it was shaded, you would be suspicious, because you know that the chance of this happening randomly is only 5 percent. If you picked two shaded shafts in a row

Exhibit 11-5. Histogram of 1,000 shafts with 5 percent shaded

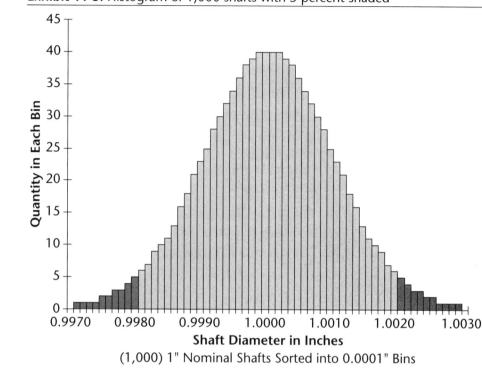

Shaft Diameter in Inches

(1,000) 1" Nominal Shafts Sorted into 0.0001" Bins

(assuming that you had put the first shaft back, mixed the shafts, and then randomly picked the second shaft), then you would *really* wonder, since you know that the chance of randomly picking two shaded shafts in this manner is only $0.05 \times 0.05 = 0.0025$, or only 0.25 percent! From this limited sample, you would suspect that the whole shaft population was more than 5 percent shaded.

Exhibit 11-6. Random shafts with 5 percent shaded

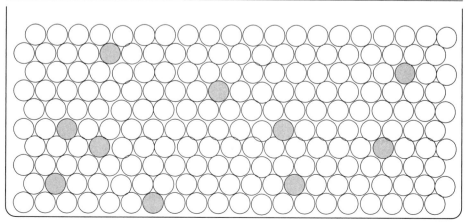

Z VALUE

The standardized normal distribution table (Exhibit 11-7) is one way to get probability values to use on any normal process or set of data. The Z in the table refers to the number of sigma to the right of the center. The probabilities refer to the area to the right of the Z point.

Exhibit 11-7. Standardized normal distribution table

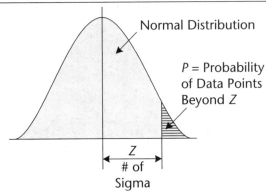

Z	P	Z	P	Z	P	Z	P
0.00	0.5000	0.05	0.4801	0.10	0.4602	0.15	0.4404
0.20	0.4207	0.25	0.4013	0.30	0.3821	0.35	0.3632
0.40	0.3446	0.45	0.3264	0.50	0.3085	0.55	0.2912
0.60	0.2743	0.65	0.2578	0.70	0.2420	0.75	0.2266
0.80	0.2119	0.85	0.1977	0.90	0.1841	0.95	0.1711
1.00	**0.1587**	1.05	0.1469	1.10	0.1357	1.15	0.1251
1.20	0.1151	1.25	0.1056	1.30	0.09680	1.35	0.08851
1.40	0.08076	1.45	0.07353	1.50	0.06681	1.55	0.06057
1.60	0.05480	1.65	0.04947	1.70	0.04457	1.75	0.04006
1.80	0.03593	1.85	0.03216	1.90	0.02872	1.95	0.02559
2.00	**0.02275**	2.05	0.02018	2.10	0.01786	2.15	0.01578
2.20	0.01390	2.25	0.01222	2.30	0.01072	2.35	0.009387
2.40	0.08198	2.45	0.007143	2.50	0.006210	2.55	0.005386
2.60	0.004661	2.65	0.004025	2.70	0.003467	2.75	0.002980
2.80	0.002555	2.85	0.002186	2.90	0.001866	2.95	0.001589
3.00	**0.001350**	3.05	0.001144	3.10	0.0009677	3.15	0.0008164
3.20	0.0006872	3.25	0.0005771	3.30	0.0004835	3.35	0.0004041
3.40	0.0003370	3.45	0.0002803	3.50	0.0002327	3.55	0.0001927
3.60	0.0001591	3.65	0.0001312	3.70	0.0001078	3.75	0.00008844
3.80	0.00007237	3.85	0.00005908	3.90	0.00004812	3.95	0.00003909
4.00	0.00003169	4.05	0.00002562	4.10	0.00002067	4.15	0.00001663
4.20	0.00001335	4.25	0.00001070	4.30	0.00000855	4.35	0.00000681
4.40	0.00000542	4.45	0.00000430	4.50	0.00000340	4.55	0.00000268
4.60	0.00000211	4.65	0.00000166	4.70	0.00000130	4.75	0.00000102
4.80	0.00000079	4.85	0.00000062	4.90	0.00000048	4.95	0.00000037

Be aware that some tables (and computer programs) use different reference points, so examine tables and computer programs carefully before you use them. Satisfy yourself that you can find data points on the standardized normal distribution table (Exhibit 11-7) relating to the previous shaft histogram with 0.0001″ bins (Exhibit 11-5).

So that there is no confusion reading this chart, let's be sure that it agrees with our reference number of 2/3 (68 percent) of data points being within ±1 sigma. Looking at the table, with a $Z = 1.00$ (which means a sigma of 1), we get $P = 0.1587$, or approximately 0.16. This is illustrated in Exhibits 11-8 and 11-9. Exhibit 11-8 shows the area on the right side of the curve.

Exhibit 11-8. Normal distribution

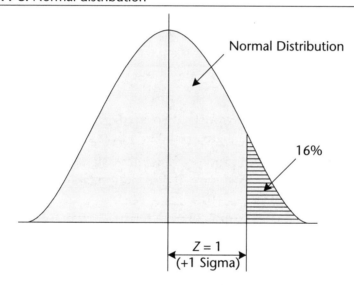

Since the left side is a mirror image of the right, this means that what we see in Exhibit 11-9 is also true.

Given that the area under the curve always equals 1 (the sum of all the probabilities equals 1), we know that the lighter area under the curve = 1 – the shaded tails. This is 1 – (16 percent + 16 percent) = 1 – 32 percent = 68 percent. This confirms our reference number of 68 percent (or 2/3, which is easy to remember).

Exhibit 11-9. Mirror image of normal distribution

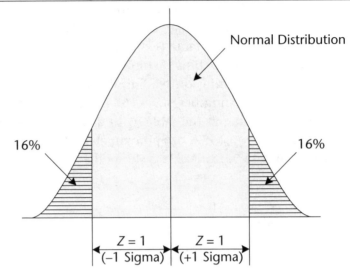

Normal Distribution

16%   16%

$Z = 1$ $Z = 1$
(−1 Sigma)   (+1 Sigma)

Problem 1

In the shaft process previously discussed, what is the probability of finding a shaft at least 2 sigma (0.0020″) over 1.000″ in diameter?

The Z value is an indication of how many sigma, so in this case, $Z = 2$. Looking at the standardized normal distribution table (Exhibit 11-7), when $Z = 2$, $P = 0.02275$.

Answer: $P = 0.02275$, or 2.28 percent.

Problem 2

What is the probability of finding a shaft that is not greater than 1.002″?

We first must realize that 1.002″ is 2 sigma above nominal (since sigma = 0.001″), so $Z = 2$. Using the standardized normal distribution table (Exhibit 11-7) to get the probability, looking at $Z = 2$, we see that $P = 0.02275$.

Looking at the normal distribution curve at the top of the table (Exhibit 11-7), we can see that this P is the probability of being *greater than* 1.002″. Since we want the probability of being *not greater than* 1.002″, we must subtract 0.02275 from 1.0000. Again, we know that we can do this because the total area under the curve, which represents all probabilities, = 1. So, $1 − 0.02275 = 0.97725$.

Answer: $P = 0.97725$, or 97.725 percent.

Assuming Normal Distribution

Use plotted data to visually determine whether the data are normally distributed. When in doubt, plot more data. Unless the data are *dramatically* nonsymmetrical (data extremely off to one side) or *dramatically* bimodal (two lobes), assume a normal distribution. The data must *clearly* show a different distribution; if they do not, we assume it is normal. There are mathematical formulas to test whether data are normal, but optical inspection of the plotted data is generally sufficient. As you will see later, having a normal distribution allows you to use the absolute probability values in the standardized distribution table directly. However, if the distributions are *not* normal, the table values can still be used for comparison purposes.

Chapter 14 will show that, as long as distributions have similar plots, they can be compared with each other even if their distributions are not normal.

For reference, Exhibit 11-10 shows a histogram example that I would consider borderline normal in that we can still use the standardized distribution table (Exhibit 11-7). The distribution, although not perfectly bell-shaped, is not skewed enough to be a concern.

The plot in Exhibit 11-10 is based on 48 data points. When in doubt, you can always plot more data.

Exhibit 11-10. Histogram of a borderline normal distribution

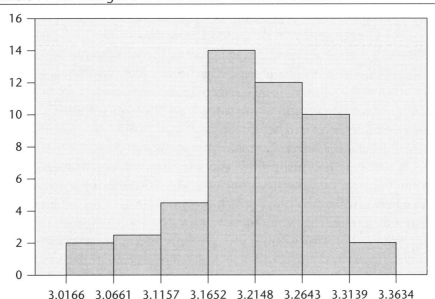

As you will see in Chapter 14, most of the work we do in Six Sigma does not require a perfectly normal distribution, since we are generally looking for relative change.

CASE STUDY: INCURRING GREAT COSTS RATHER THAN PLOTTING

A high-volume consumer product, with more than $50 million per year in sales, had a historical increase of sales of 3 percent a year. This growth rate was expected to slowly decrease because of competing products having a longer life. To everyone's surprise, the sales actually increased 13 percent within a year.

Many people had different theories as to why this happened, but no one had any supporting data. The theories for the increased sales ranged from a sudden need by consumers for more of this specific product to an excellent marketing campaign. Plans to expand production facilities to support these sales were started, since the company was having great difficulty meeting this unexpected demand.

The marketing of this product was divided into two units, one that handled large-volume outlets and another that handled mom-and-pop small stores. After more than a year of the increased sales and almost $500,000 spent on expansion design, someone noticed that the increase in sales had hit the large-volume outlets well before it had affected the low-volume outlets. Since both kinds of outlets served similar customers, this was mysterious. Finally, someone attributed the cause of the increased sales to a design change that had been implemented in the product sometime earlier. This design change had inadvertently reduced average product life 10 percent. Since the large-volume outlets used just-in-time inventories, their customers experienced the effect of the shorter life far earlier than the small outlets, whose inventory usually covered many months of sales. Thus, the high-volume outlets felt the increased sales level well before the small outlets did. The design change was reversed—and the unexpected sales increase disappeared.

If someone had just plotted the sales from these two marketing units when the sales increase first appeared, she would have spotted the difference between the two plots, which previously had shown the same 3 percent growth rate. This would have triggered a more extensive cause analysis one year earlier. This would have prevented $5 million in excessive costs to consumers and a loss of some customers. Although the supplier had a short-term windfall from the increased sales, it lost a great deal of long-term business because it could not supply the product on a timely basis.

This case study shows how people are quick to react with solutions (expand the facilities) despite their great cost, but will spend little time on plotting data and truly understanding a root cause.

Normal Distribution Symmetry   TIP

Remembering that each half of the normal distribution curve is a mirror image of the other, we can use data given for the plus side to solve problems related to both sides.

Problem 3

In the shaft process previously discussed, what is the probability of getting a shaft that is below 0.9978″ in diameter?

This shaft diameter is 0.0022″ below nominal (1 − 0.9978″). Since sigma = 0.0010″, this is 2.2 sigma below nominal, so $Z = 2.2$. Looking at the standardized normal distribution table (Exhibit 11-7), for $Z = 2.2$, we see that $P = 0.0139$. Thus, 1.39 percent of the data points would occur above a positive 2.2 sigma. Since the negative side of the probability table is a mirror image of the positive side, this probability also applies to a negative 2.2 sigma.

Answer: $P = 0.0139$, or 1.39 percent.

No Values Occur Exactly at a Z Point   TIP

In using a standardized normal distribution curve, all values are assumed to be above or below a Z point. For example, if you wanted to know what percent of values are "above $Z = 2$," it would be the same as the percent of values "at or above $Z = 2$."

For simplicity, the previous shaft data had a sigma = 0.0010″. This was to make calculations and understanding easier. Usually the sigma doesn't correlate with the bin edges, nor is it such an even number. This in no way changes the logic or diminishes the value of the standardized normal distribution table (Exhibit 11-7), as illustrated in the next problem.

Problem 4

Using the shaft example, let's assume that the customer has complained that the amount of variation in the shafts is causing him process problems. The customer is especially critical of shafts that are less than 0.9980″ and greater than 1.0020″ (more than 0.0020″ from nominal). In response, the lathe is overhauled. Upon taking another

1,000 measurements, it is determined that the average has stayed at 1.0000", but the sigma has been reduced from 0.0010" to 0.0007".

The reduced sigma means that the variation among shafts is less than it was before the overhaul. We want to communicate to the customer what improvement he can expect in future shipments—specifically, what reduction he will see in shafts more than 0.0020" above or below the nominal 1.0000" diameter.

Before the overhaul (Problem 1, sigma = 0.0010"), we found that the probability of finding a shaft at least 0.0020" above 1.0000" in diameter was 0.02275. Given that the two sides of the curve are mirror images, we doubled that number to calculate the chances of being at least 0.0020" ± nominal.

$$P = 0.02275 \times 2 = 0.0455, \text{ or } 4.55 \text{ percent } (P \textit{ before} \text{ the overhaul})$$

We must now calculate the P with the new reduced sigma (0.0007"). First, we must see how many sigma "fit" between the nominal value and 0.0020". We use the plus side first, since that is the data given to us in the standardized normal distribution table (Exhibit 11-7).

0.0020"/0.0007" = 2.86 sigma fit! This gives us the Z to use in the standardized normal distribution table (Exhibit 11-7).

Using the standardized normal distribution table (Exhibit 11-7), looking at Z = 2.85 (the closest data point in the table), the P value we read from the table is 0.002186. So, 0.2186 percent of the shafts will be at least 0.0020" above nominal. We double this to include those that are at least 0.0020" below 1.0000" diameter.

$$2 \times 0.002186 = 0.004372$$

The total P is 0.004372, or 0.4372 percent (P *after* the overhaul).

Answer: Since the process had been making 4.55 percent at 0.0020" above or below 1.0000" and it is now making 0.4372 percent, the customer can expect to see only 9.6 percent (0.437/4.55) of the former problem shafts.

Problem 5

Let's change the previous problem again to make it even more "real." After the overhaul, the lathe sigma is reduced to 0.0007" (same as before), but the average shaft diameter is now 1.0005". The process plot is still normal. Will the customer be receiving fewer problem shafts than before the overhaul?

Since the process average is no longer centered at the 1.0000" nominal, the amount of product outside the 0.9980"-to-1.0020" target range is different for the large diameters from what it is for the small diameters, so we need to calculate each independently.

First, we will calculate the P for the too-large shafts. As before, we see how many sigma (0.0007″) "fit" between the new process average (1.0005″) and the +1.0020″ upper limit.

This calculation is $(1.0020″ − 1.0005″)/0.0007″ = 2.143$ sigma fit.

Looking at the standardized normal distribution table (Exhibit 11-7), we see that the P at a Z of 2.15 (the closest value to 2.143) is 0.01578. That means that 1.578 percent of the shafts will be 1.0020″ in diameter or larger.

Looking at the too-small shafts, we do a similar calculation. First, we need to find the value for the difference between the process average and the lower end of the target range. (The process average is 1.0005″, and the lower target value is 0.9980″.) The difference is $1.0005″ − 0.9980″ = 0.0025″$.

We then see how many sigma (0.0007″) "fit": $0.0025″/0.0007″ = 3.57$ sigma. Although we must use the data on the positive end of the curve, we know that the mirror image would be identical. Looking at the P value for a Z of 3.55, we get $P = 0.0001927$. So, 0.019 percent of the shafts will be 0.9980″ or smaller.

Answer: When we add the too-large- and too-small-diameter shafts, we get 1.578 percent + 0.019 percent = 1.60 percent of the shafts will be at least 0.0020″ off the 1.0000″ nominal. Since 1.60 percent is less than the 4.45 percent that the customer was receiving before the overhaul, the customer will be receiving a better product. Note, however, that 1.60 percent is much higher than the 0.4372 percent (Problem 4) that the customer would receive if the process were centered.

This change in both the average and the sigma is not unusual in a process change. However, it is usually not difficult to get the process mean back to the target center (in this case, 1.0000″ diameter). If the process center is put back to nominal, we get the tenfold improvement that we saw in the earlier problem.

Adjusting a Process's Mean Versus Reducing Its Sigma

Normally, moving a process's mean is easier than trying to reduce its sigma. A mean change often just involves choosing the center around which the process will be run; it requires no major process change. A sigma reduction often requires a significant change in the process itself, like dramatically slowing the process or changing the equipment being used.

TIP

Note that in the previous cases, the Z values from the standardized normal distribution table (Exhibit 11-7) that were used were those closest to the calculated values of Z. There was no attempt to extrapolate or to go to another table or a computer program for greater accuracy. Either would have been possible, but if you look at the relative values obtained versus the changes being noted, the greater accuracy was not required. Often the calculation accuracy far exceeds the requirements of the output results.

Using Excel to Get Normal Distribution Values

TIP

Those who wish to use the computer to get the probability values for various values of Z can use Excel. After bringing up the Excel worksheet, click on "Formulas," click on "More functions," then "Statistical." Click on "NORM.S.DIST." When you enter a Z value along with cumulative = true, then "OK," it gives you the probability values using the left end of the distribution as the reference zero, whereas the standardized normal distribution table (Exhibit 11-7) uses the right end of the distribution as zero. To convert either one to the other, simply subtract the value from 1.

For example, if you enter $Z = 2$ with cumulative = true in the Excel NORM.S.DIST, you get a probability value of 0.97725. This is the probability of being less than the Z value. $1.00000 - 0.97725 = 0.02275$ is the probability of being greater than Z, which matches the probability given for $Z = 2$ on the standardized normal distribution table (Exhibit 11-7).

Just for information purposes, the Six Sigma process is sometimes referred to as three defects per million. If you look at the standardized normal distribution table (Exhibit 11-7), you will see that three defects per million is 4.65 sigma, not 6 sigma. The 6-sigma short-term target is tighter than 4.65 sigma because it assumed that a process drift would take place over time. If you started with a process that was 6 sigma in the short term, the goal was to have a 4.65-sigma process when the long-term drift was included.

PLOTTING DATA

There are hundreds of computer programs available that will plot data and do some degree of statistical analysis. Some of these are quite good; many are somewhat confusing. Generally, the more ambitious the program (three-dimensional plots in various colors, every type of plot imaginable, esoteric statistical analysis), the more chances of getting an output that doesn't tell the desired story. This problem is caused by incomplete or confusing directions or help screens, the user not taking the time to understand the details of the program, or even errors within the program.

CASE STUDY: NOT TESTING A GRAPH FOR REASONABLENESS

In a large corporation, there was a review of Six Sigma projects. In attendance were many black belts and green belts. The presenter was displaying what he described as "normal" data consisting of 100 individual data points, with the ±3 sigma lines shown on the graph (Exhibit 11-11).

Exhibit 11-11. Graph: normal data

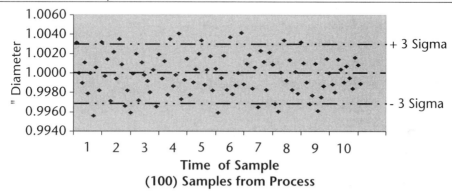

During the presentation, no one questioned this chart or the resultant conclusions. At the end of the presentation, however, one person asked how approximately 20 percent of the data points could be beyond the 3-sigma limits, since the limits were supposedly calculated from the data points displayed. (If you will recall, one of the rules of thumb is that 99.7 percent of the data points in a normal distribution are within the ±3-sigma limits.) This triggered some negative comments aimed at the questioner, inquiring whether he thought he was smarter than the person who wrote the software program. There also was general confusion, since the program used was the designated statistical program for the whole corporation. Only later did someone discover that within the program was a default that used the last 10 data points entered to calculate the 3-sigma limits.

This case study is problematic for several reasons. First, other than the questioner, no one demonstrated a basic understanding of what the 3-sigma limits meant; no one tested the graph for reasonableness. Second, it's troublesome that the default in the computer program would use only the last 10 points entered to calculate the 3-sigma limits. (You will learn later that a *minimum* of 11 points is needed to get a decent estimate of sigma, with 30 points being preferred.) Third, almost no one using this program had bothered to understand how the program worked (or its defaults), or what was the basis of its output.

Many programs are so forbidding that the user is relieved just to get an output. There is also a feeling that a graph output is some verification that the input is correct and that the output is meaningful. Many of these programs are powerful, but require care to use.

Any users of a statistical program that is not completely familiar to them should do some manual work with the data before they use the program. They should then input a very simplified set of data for which they already know the outcome. Finally, they should test the output carefully for reasonableness.

In this book, we use Excel to generate our graphs. This is *not* because Excel is the best program for graphs (it isn't), but because I'm assuming that most users of this text will have Microsoft's Excel on their computers. Details of other programs must be left to individual users.

Using Excel's Graphing Program

I will go into more detail than you probably need in case you are less Excel-oriented. If you are completely familiar with graphing with Excel, you can simply glance over this section quickly. Note that different editions of Excel can vary somewhat on detail. The following is based on the Excel 2010.

First, make sure that Excel's Data Analysis program is loaded in your computer. Bring up Microsoft Excel. On the header at the top of the screen, go to "Data." See if "Data Analysis" is one of the options available. If not, under "File," click "Options," "Add-ins," then highlight "Analysis ToolPak." Under "Manage Excel add-ins," click "GO." A pop-up screen shows add-ins available. Put a check in the box next to "Analysis ToolPak." Click "OK." Go back to a spreadsheet.

Copy the following 50 numbers into an Excel worksheet, column A.

1.0004	0.9994	1.0008	1.0013
0.9991	0.9978	1.0005	0.9984
1.0001	0.9982	1	0.9996
0.9995	1.0015	0.9996	1.0009
1.0012	1.001	0.9997	0.9998
0.9998	1.0016	0.9985	0.9998
0.9986	0.999	0.9991	0.9989
0.9999	0.9989	1.0004	0.9993
1	1.0001	0.9993	1.0003
0.9996	0.9972	1.0006	1.0019
1.0002	1.0009	0.9991	0.9996
0.9988	1.0024	1.0002	
1.0009	0.9996	1.0014	

These numbers represent 50 shaft diameter readings that we may expect from the previously discussed shaft process. After copying these numbers, highlight them. Go into "Data" on the toolbar, and click on the AZ down arrow opposite "Sort." After you have ordered these numbers, the top number will be 0.9972 and the bottom number will be 1.0024.

In column B, row 1, enter the formula "= (bottom or maximum number) – (top or minimum number)," which in this case will be "= A50 – A1." This will give the difference between the largest and smallest shaft diameter, which will be 0.0052.

In that same column B, second row, insert the formula "= 1.02 ∗ B1 / 7." This gives us bin sizes for seven bins. (See the following tip for calculating the number of bins.) The use of "1.02" makes the total bin widths slightly wider than the data range. If we have more data, we can use more bins by changing the denominator from 7 to a higher bin quantity. With seven bins, the bin width shown in B2 will be 0.0007577.

Now we have to show the specific bin edges. In C1, insert the formula "= A1 – 0.01 ∗ B1." In this example, it will put a value of 0.99715 in C1. This gives the left bin edge, which is slightly less than the minimum data value. Then, in C2, insert the formula "= C1 + B2." This determines that the next bin edge will be the number in C1 plus the bin width. C2 in this example will then be 0.99791. The $s in this case "freeze" the bin width B2 for use in the next steps.

Highlight C2, go to "Home" in the main tool bar, "Copy" (icon next to Paste), then highlight C3 through C8 (you would highlight more if you had more bins), then "Paste." This gives you the edge values for each of the remaining bins. In this example, C8 should show the bin value 1.00245, the right edge of the last bin. This value is slightly higher than the maximum data number.

Now, go to "Data" in the top header, then to "Data Analysis > Histogram." "OK." The cursor will be on the "Input Range." Highlight the data in column A. Click on the second box ("Bin Range"). Highlight all the data in column C. The options "New Worksheet Ply" and "Chart Output" should be chosen, and then hit "OK."

The Histogram will come up. The histogram bars should show the same values as in Exhibit 11-12. I modified Exhibit 11-12 to bring the bars together and changed the *x* axis labels somewhat, and you can modify your chart as desired. However, in this book, we are more interested in the general shape of the histogram rather than making the histogram look attractive.

Exhibit 11-12. Histogram of sample data

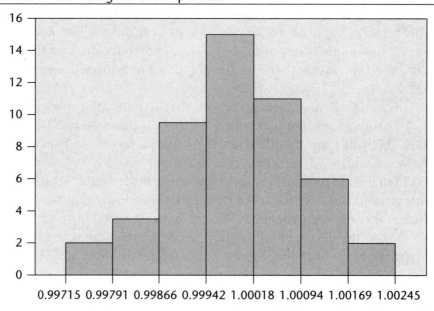

Rule of Thumb for the Number of Bins to Use in a Histogram

$$\# \text{ of Bins} = \sqrt{\# \text{ of Data Points}}$$

Note: This is only a general guideline. Feel free to experiment.

Additional Practice Problems

Use the following information and the standardized normal distribution table (Exhibit 11-7) on the following additional problems.

An insurance company has plotted hospital bills for delivering a baby when there are no complications and has found that the distribution is normal. In a specific city, the average delivery cost is $3,020, with a standard deviation of $280.

Problem 6
What is the probability of the hospital bill for a normal delivery in this city being greater than $3,380?

Problem 7

What is the range (high and low) of hospital bills that 95 percent of people would expect for a normal delivery in this city?

Problem 8

The insurance company has found that delivery costs in this city are greater than those in other similar cities. Should the insurance company emphasize reducing the average cost or the variation in costs within this city?

Problem 9

The insurance company makes a concerted effort to reduce delivery costs in this city. After a year, it finds that it has reduced the average cost from $3,020 to $2,910. The sigma stayed the same at $280. What percent reduction can it expect in the number of bills over $3,200?

Problem 10

As in the previous problem, the insurance company reduces the average delivery cost from $3,020 to $2,910, but the standard deviation goes from $280 to $305. What percent reduction can it expect in the number of bills over $3,200?

Problem 11

Using Excel and the following 48 numbers, create a histogram.

3.24	3.25	3.25	3.29
3.21	3.36	3.26	3.28
3.29	3.28	3.3	3.18
3.28	3.18	3.18	3.22
3.32	3.21	3.19	3.24
3.21	3.25	3.1	3.17
3.25	3.18	3.11	3.16
3.13	3.29	3.18	3.06
3.14	3.1	3.14	3.31
3.26	3.22	3.19	3.31
3.22	3.02	3.21	3.27
3.2	3.19	3.12	3.25

Solutions to Additional Practice Problems

An insurance company has plotted hospital bills for delivering a baby when there are no complications, and has found that the distribution is normal. In a specific city, the average delivery cost is $3,020, with a standard deviation of $280.

Problem 6

What is the probability of the hospital bill for a normal delivery in this city being greater than $3,380?

$$\$3,380 - \$3,020 = \$360$$
$$\$360/\$280 = 1.2857, \text{ so } Z = 1.286$$

Looking at the standardized normal distribution table (Exhibit 11-7), for $Z = 1.30$ (the closest to 1.286), $P = 0.09680$.

So, the chance of getting a bill greater than $3,380 is approximately 9.7 percent.

Problem 7

What is the range (high and low) of hospital bills that 95 percent of people would expect for a normal delivery in this city?

We first solve for the high end of the range, since the standardized normal distribution table (Exhibit 11-7) shows that end.

$95\%/2 = 47.5\%$ of the bills are in the upper half.

$50.0\% - 47.5\% = 2.5\%$ is the probability of a bill that is higher than expected.

Looking within the standardized normal distribution table (Exhibit 11-7) for $P = 0.025$ (2.5%) gives $Z = 1.95$ (the closest Z).

Multiply the Z times the sigma value to calculate how much higher the upper end of the range is than the average:

$$1.95 \times \$280 = \$546$$

Adding this to the mean gives us the high end of the range:

$$\$3,020 + \$546 = \$3,566$$

Since we know that a normal distribution is symmetrical, we know that the low end of the range will be an equal amount below the average:

$$\$3,020 - \$546 = \$2,474$$

So, 95 percent of the people in this city can expect to pay between $2,474 and $3,566 for a normal delivery.

Problem 8

The insurance company has found that delivery costs in this city are greater than those in other similar cities. Should the insurance company emphasize reducing the average cost or the variation in costs within this city?

In general, the average is easier to change than the variation. The insurance company has a better chance of reducing its average costs (by putting out general guidelines, encouraging the use of generic drugs, and so on) than of reducing the variation among all doctors and hospitals. Of course, some reduction in variation may also come as a secondary benefit.

Problem 9

The insurance company makes a concerted effort to reduce delivery costs in this city. After a year, it finds that it has reduced the average cost from $3,020 to $2,910. The sigma stayed the same at $280. What percent reduction can it expect in the number of bills over $3,200?

First let's calculate the number of bills over $3,200 at the initial $3,020 average cost, with the sigma of $280. We need to get the Z value:

$$Z = (\$3,200 - \$3,020)/\$280 = 0.643$$

From the standardized normal distribution table (Exhibit 11-7), $P = 0.2578$ (25.78 percent).

So, at the initial $3,020 average, 25.78 percent of the bills were over $3,200.

Now calculate the number of bills over $3,200 at the $2,910 average cost, with the sigma of $280. We need to get the Z value:

$$Z = (\$3,200 - \$2,910)/\$280 = 1.0357$$

From the standardized normal distribution table (Exhibit 11-7), $P = 0.1469$ (14.69 percent).

So, at the lower $2,910 average, 14.69 percent of the bills are over $3,200.

The difference is 25.78 percent − 14.69 percent = 11.09 percent. 11.09 percent/25.78 percent = 0.430.

So, the hospital will see an approximately 43 percent reduction in the number of bills over $3,200.

Problem 10

As in the previous problem, the insurance company reduces the average delivery cost from $3,020 to $2,910, but the standard deviation goes from $280 to $305. What percent reduction can it expect in the number of bills over $3,200?

Again, first calculate the number of bills over $3,200 at the initial $3,020 average cost, with the sigma of $280. We need to get the Z value:

$$Z = (\$3,200 - \$3,020)/\$280 = 0.643$$

From the standardized normal distribution table (Exhibit 11-7), $P = 0.2578$ (25.78 percent).

So, at the initial $3,020 average, 25.78 percent of the bills were over $3,200.

Now calculate the number of bills over $3,200 at the $2,910 average cost, with the sigma of $305. We need to get the Z value:

$$Z = (\$3,200 - \$2,910)/\$305 = 0.951$$

From the standardized normal distribution table (Exhibit 11-7), $P = 0.1711$ (17.11 percent).

So, at the lower $2,910 average, 17.11 percent of the bills were over $3,200.

The difference is 25.78 percent – 17.11 percent = 8.67 percent. 8.67 percent/25.78 percent = 0.336.

So, the insurance company will see an approximately 34 percent reduction in the number of bills over $3,200.

Problem 11

Using Excel and the following 48 numbers, create a histogram.

3.24	3.25	3.25	3.29
3.21	3.36	3.26	3.28
3.29	3.28	3.3	3.18
3.28	3.18	3.18	3.22
3.32	3.21	3.19	3.24
3.21	3.25	3.1	3.17
3.25	3.18	3.11	3.16
3.13	3.29	3.18	3.06
3.14	3.1	3.14	3.31
3.26	3.22	3.19	3.31
3.22	3.02	3.21	3.27
3.2	3.19	3.12	3.25

After copying these 48 numbers, highlight them. Go into "Data" on the toolbar, and click on the AZ down arrow opposite "Sort." After you have ordered these numbers, the top number will be 3.02 and the bottom number will be 3.36.

In column B, row 1, enter the formula "= (bottom or maximum number) – (top or minimum number)," which in this case will be "= A48 – A1." This will give the difference between the largest and smallest shaft diameter, which will be 0.34.

In that same column B, second row, insert the formula "= 1.02 * B1 / 7." This gives us bin sizes for seven bins. With seven bins, the bin width shown in B2 will be 0.049543.

Now we have to show the specific bin edges. In C1, insert the formula "= A1 – 0.01 * B1." In this example, it will put a value of 3.0166 in C1. This gives the left bin edge, which is slightly less than the minimum data value. Then, in C2, insert the formula "$= C1 + \$B\$2$." This determines that the next bin edge will be the number in C1 plus the bin width. C2 in this example will then be 3.0661. The $s in this case "freeze" the bin width B2 for use in the next steps.

Highlight C2, go to "Home" in the main toolbar, "Copy" (icon next to Paste), then highlight C3 through C8 (you would highlight more if you had more bins), then "Paste." This gives you the edge values for each of the remaining bins. In this example, C8 should show the bin value 3.3634, the right edge of the last bin. This value is slightly higher than the maximum data number.

Now, go to "Data" in the top header, then to "Data Analysis > Histogram." "OK." The cursor will be on the "Input Range." Highlight the data in column A. Click on the second box ("Bin Range"). Highlight all the data in column C. The options "New Worksheet Ply" and "Chart Output" should be chosen, and then hit "OK."

The Histogram will come up. The histogram bars should show the same values as in Exhibit 11-13. I modified Exhibit 11-13 to bring the bars together and changed the x axis labels somewhat, and you can modify your chart as desired. However, in this book, we are more interested in the general shape of the histogram rather than making the histogram look attractive.

Exhibit 11-13. Histogram for Problem 11

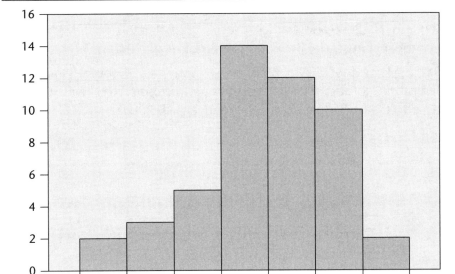

WHAT WE HAVE LEARNED IN CHAPTER 11

1. Plotting data is a necessary step in implementing many of the Six Sigma tools.
2. Data needed for histograms and standardized normal distribution table analysis are often readily available.
3. Using histograms to compare supposedly similar areas or year-to-year performance helps spot unexpected differences and areas of opportunity. Using the standardized normal distribution table (Exhibit 11-7) to evaluate data on a normal distribution or to compare two processes with similarly shaped histograms can often help to quantify a problem.
4. Excel can be used to make histograms or to get normal distribution values.
5. A graphing program may be powerful, but the user needs to fully understand it.
6. You can do real Six Sigma work by using histograms and the normal distribution table.

RELATED READING

Mark J. Kiemele, Stephen R. Schmidt, and Ronald J. Berdine, *Basic Statistics: Tools for Continuous Improvement*, 4th ed. (Colorado Springs, CO: Air Academy Press, 1997).

John Walkenbach, *Excel Charts* (with CD-ROM) (Indianapolis, IN: John Wiley & Sons, 2003).

Kenneth N. Berk and Patrick Carey, *Data Analysis with Microsoft Excel* (Southbank, Australia; Belmont, CA: Brooks/Cole, 2004).

David M. Levine, David Stephan, Timothy C. Krehbiel, and Mark L. Berenson, *Statistics for Managers, Using Microsoft Excel* (with CD-ROM), 4th ed. (Upper Saddle River, NJ: Prentice Hall, 2004).

PART IV

Six Sigma Tools to Test for Statistically Significant Change

CHAPTER 12

Testing for Statistically Significant Change Using Variables Data

W hat you will learn in this chapter is how to use limited samples of variables (measurable) data to make judgments concerning whether a population or process has changed. We will be comparing samples with the population and samples with other samples. We first want to determine the minimum sample size required. This is very important, because taking too small a sample can cause invalid estimates, but excessive samples are costly. The material in this chapter is used in the Define, Measure, Analyze, Improve, and Control steps in the DMAIC process.

Changes in Real Processes, Variables Data

Manufacturing. Compare dimensional samples from two similar production lines or compare samples from different shifts on one line to look for significant differences.
Verify the results of a test by comparing a sample against the population or against another sample.

Sales. Compare samples of sales dollars generated by different salespeople, using samples from many random days.

Marketing. Compare samples of advertising dollars spent in a city over a period of time versus the sales generated in that city to see whether advertising dollars make a significant difference in sales.

Accounting and software development. Sample error rates and look for differences among people or departments doing similarly complex work, using samples from many individual days.

Receivables. Sample delinquent receivables to check for a correlation with company size or D&B rating. Use data from many individual days.

Insurance. Sample costs among treatment centers on similar procedures, using data from many individual days.

In the chapter on probability, we learned to judge whether an event was random or due to an assignable cause. In all cases, we knew the specific odds of the random event, like 0.5 on a coin flip or 1/6 on rolling a die. If an event were nonrandom, this hinted at either a problem or an opportunity.

In the chapter on data plots and distributions, we plotted a large quantity of data from a sample or population to see trends or changes. We then used the plot to see the distribution of data. We used a standardized normal distribution table to do further analysis. As in the probability chapter, once we had an identified distribution, we used this knowledge to judge whether an event was random or due to an assignable cause.

Variables Data

Variables data are measurable, generally in decimal form. Theoretically, if you looked at enough decimal places, you would find that no two values are exactly the same. *Continuous* data is another term used for this type of data.

DEFINITION

Proportional Data

Proportional data are based on attribute inputs, such as "good" or "bad," "yes" or "no," and other such distinctions. Examples are the proportion of defects in a process, the proportion of "yes" votes for a candidate, and the proportion of students failing a test.

DEFINITION

Sample Size Rule of Thumb on Variables Data

To calculate sigma on variables data, a rule-of-thumb minimum sample size *n* of 11 is needed, with a preferable sample size *n* of 30 or more.

TIP

> To calculate an average, the sample size *n* can be as low as 6. But, since we normally must calculate sigma at the same time, the minimum *n* of 11 and preferable *n* of 30 are the standard.
>
> This rule of thumb should be used only when it is not possible to calculate a specific minimum sample size. This would be when we don't know the population sigma or are not sure of the accuracy we need.
>
> Many quality plans use less than the rule-of-thumb sample size because of sampling cost issues. However, anyone implementing such a plan should be aware that he may be sacrificing some level of confidence in the results.

CASE STUDY: OBSCENE SCRAP CALL

In a process that ran 24 hours per day, an inspector took product readings continuously. Every time seven pieces had been measured, a computer program calculated an average and sigma from these seven pieces. Using that average and sigma, the computer projected the percentage of the product that was outside specifications. The results of these calculations were then displayed in the operator's booth, which was at the other end of the manufacturing plant. If the projected percentage of the product that was outside specifications was too high, the product was put into "scrap" and remained there until the calculations on another seven-piece sample showed an acceptable defect projection.

The scrap decision was communicated to the machine operator, who was supposed to make adjustments to bring the product back to acceptable levels. The operators had learned from experience, however, that these scrap decisions were sometimes invalid and that the next sample would often be acceptable without any process adjustments being made. Therefore, the operator would delay adjusting the process until several scrap calls occurred in a row.

The result was that scrap time was excessive, bad product was shipped (the erroneous scrap calls also missed some bad product), and the plant production people had no faith in the quality system. The machine operators had even coined an obscene phrase for the basis of the scrap decision. The first half of this phrase was "sigma," which I am not sure they fully understood. But they certainly knew what the last half of the coined phrase meant!

According to the earlier tip on sample size, the rule of thumb for the minimum sample size for a valid sigma estimate is 11. The quality system in this example was using a sample size of 7! The quality manager was very hesitant to increase the sample size, because he thought the system reaction time would then be too slow. It took much convincing to get him to even try a larger sample size.

When the sample size was increased to 15, the number of erroneous scrap calls was substantially reduced (this was verified by data taken before and after the change), the outgoing quality was improved, scrap time was reduced, and faith in the quality system was restored. The reaction time of the system never became an issue, since having valid scrap calls was far more important. The scrap time reduction alone was worth more than $100,000 per year, and there was negligible cost for implementation.

Almost any formal or informal quality system has a sample size that is used to make decisions and to give feedback to the people involved. Even where there are no formal rules, someone makes a judgment concerning how many items should be reviewed before making a determination. This is true in reviewing office staff output, incoming product, medical errors, accounting errors, programming mistakes, and so on. If too many examples are required, a decision is delayed and poor performance is missed. If judgment is made on too small a sample, then erroneous calls are often made.

DATA DISCRIMINATION

Some data need interpretation to determine whether they can be analyzed as variables. The following example shows how this determination can be made.

The following list are individual ages collected from a group of people aged 50 to 55. Assume that these data are going to be used in a study of the incidence of a disease.

Ages

53.2	55.0	54.2
54.1	52.7	53.4
50.2	54.5	52.3
54.3	54.8	53.4
51.3		

Since the data were collected in units equal to tenths of a year, there is good discrimination of the data relative to the five-year study period. Good discrimination means that the data steps are small compared to the five-year range being studied. These data have measurement steps of tenths of a year because the data were collected and displayed to one-decimal-point resolution.

Rule of Thumb on Variables Data Discrimination

For variables data, the measurement resolution should be sufficient to give at least 10 steps within the range of interest. For example, if data are collected on a part with a 0.050" tolerance, the data measurement steps should be no larger than 0.005" if we are to be able to analyze those data as variables.

In the example on ages, because the ages are displayed to the first decimal place, there are 50 discrimination steps in the five-year study period (5.0 divided by 0.1), which is greater than the 10-step minimum rule of thumb.

Let's suppose that the age data had been collected as ages 53, 54, 50, 54, 51, 55, 53, 55, 55, 54, 53, 52, and 53.

Because of the decision to collect the data as discrete whole numbers, these data now have to be analyzed as proportions. With these broad steps, there are only five discrete steps within the five-year study period (5 divided by 1), which is below the 10-discrimination-step minimum for variables data. To analyze these data as proportions, for example, we could study the proportion of people 52 years of age or younger or the proportion of people 53 years of age.

The original age data could also have been collected as attributes, where each person from the given group of people was asked whether she was 53 or older. The collected data would then have looked like the following:

Is Your Age 53 or Older?

Yes	Yes	Yes
Yes	No	No
No	Yes	Yes
Yes	Yes	
No	Yes	

These are *attribute data*, which can be studied only as proportional data. For example, the proportion of yes or no answers could be calculated directly from the data, and then the resultant proportion could be used in an analysis.

What we saw in these examples is that data collected on an event can often be either variables or proportional, depending on the manner in which they are collected and their resolution or discrimination. Attribute data, however, are always treated as proportional (after determining a ratio).

There are some data that are numerical, but are available only at discrete levels. Shoe size is an example. By definition, the steps are discrete, with no information available for values between the shoe sizes. Discrete data are often treated as proportional because the steps may limit the discrimination relative to the range of interest.

Let's look at another example, this time using shaft dimensions. Assume that a shaft has a maximum allowable diameter of 1.020" and a minimum allowable diameter of 0.080", which is a tolerance of 0.040". Here is the first set of data for shaft diameter in inches:

1.008	0.997	1.003
0.982	1.000	0.991
0.996	1.009	1.009
1.017	1.002	1.014

Since the dimensions are to the third decimal place, these measurement "steps" are small compared to the need (the 0.040" tolerance). Since there are 40 discrimination steps (0.040" divided by 0.001"), these data can be analyzed as variables data.

Suppose, however, that the same shafts were measured as follows:

1.01	1.00	1.00
0.98	1.00	0.99
1.00	1.01	1.01
1.02	1.00	1.01

Since the accuracy of the numbers is now only to two decimal places, there are only four steps within the 0.040" tolerance (0.040" divided by 0.010"), which is below the rule-of-thumb 10-discrimination-step minimum. We must therefore analyze these data as proportions. For example, we could calculate the proportion of shafts that are below 1.00" and use that information in an analysis.

Similarly, the same data could have been collected as attributes, checking whether the shafts were below 1.000" in diameter, yes or no, determined by a go/no-go gauge set at 0.9999". The collected data would then look like the following list.

Is the Shaft Below 1.000" in Diameter?

No	Yes	No
Yes	No	Yes
Yes	No	No
No	No	No

These are attribute data, which can be analyzed only as proportional data. The proportion of yes or no answers can be calculated and then used in an analysis.

Again, what we saw in this example is that data collected on the same event can be either variables or proportional, depending on the manner in which items were measured and the discrimination. Once you understand this concept, it is quite simple. Since at some point, all measurements become stepped as a result of measurement limitations, these steps must be reasonably small compared to the range of interest if we wish to use the data as variables. Any data that do not qualify as variables must be treated as proportional.

Labeling Averages and Standard Deviations

The average of a population is labeled $\bar{X}$, whereas the averages of samples are labeled $\bar{x}$.

Similarly, the standard deviation (sigma) of the population is labeled S and the sample standard deviations are labeled s.

We will use $\bar{X}$, S, $\bar{x}$, and s in this book rather than the Greek letters used in some texts. Both are in use and acceptable. The feeling of the author is that the Greek letters are sometimes intimidating and give no real benefit except to appear esoteric. Statistics has enough of a stigma without unnecessary intimidation.

Population vs. Sample

We seldom have *all* the data on a population, but we make an estimate about the population based on large or multiple samples.

TIP

ESTIMATING A POPULATION AVERAGE $\bar{X}$ AND SIGMA S

Here are two ways of determining the population $\bar{X}$ and S. You can estimate a population average $\bar{X}$ and sigma S from the $\bar{x}$ and s of a large sample with a minimum sample size of $n = 30$. Assume that the average and sigma for the population are the same as those for the large sample.

You can also estimate the population sigma if you have two or more samples of similar size n. The formula is shown in the box.

Estimating Population $\bar{X}$ and S from Multiple Samples of Similar Size n

$$\bar{X} = \frac{\bar{x}_1 + \bar{x}_2}{2}$$

$\bar{X}$ is the population average
$\bar{x}_1$ is the average from the first sample
$\bar{x}_2$ is the average from the second sample

FORMULA

$$S = \sqrt{\frac{s_1^2 + s_2^2}{2}}$$

S is the population sigma
s_1 is the sigma of the first sample
s_2 is the sigma of the second sample

If you have three or more samples, modify the formulas accordingly, with more sample sigma or averages in the numerator and dividing by the total number of samples.

Example: If you have $s_1 = 16$ and $s_2 = 12$, then

$$S = \sqrt{\frac{s_1^2 + s_2^2}{2}} = \sqrt{\frac{16^2 + 12^2}{2}} = 14.14$$

If the sample sizes are substantially different, use the s from the larger sample for the estimate of S, since the confidence related to a small sample is suspect. You can also use a computer program that will compensate for different sample sizes, but this is not normally necessary.

Problem 1
You have data from two samples taken from a stable process, with no other knowledge of the process population.

Sample 1	Sample 2
$n = 16$	$n = 36$
$\bar{x} = 14.96$	$\bar{x} = 15.03$
$s = 1.52$	$s = 1.48$

What is the estimate for the population average $\bar{X}$ and sigma S?

Answer: Since the sample sizes n are quite different from each other, use the larger sample. Also, since the sample size of sample 2 is greater than 30, we feel comfortable that it is a reasonable estimate. Thus, the estimate for the population is:

$\bar{X} = 15.03$
$S = 1.48$

Problem 2
You have data from three samples taken from a stable process, with no other knowledge of the process population.

Sample 1	Sample 2	Sample 3
$n = 16$	$n = 18$	$n = 15$
$\bar{x} = 14.96$	$\bar{x} = 15.05$	$\bar{x} = 15.04$
$s_1 = 1.52$	$s_2 = 1.49$	$s_3 = 1.53$

What is the estimate for the population average $\bar{X}$ and sigma S?

Answer: Since all three samples have a similar sample size n, we will use the formulas given previously to calculate our estimate for the population.

$$\bar{X} = \frac{\bar{x}_1 + \bar{x}_2 + \bar{x}_3}{3}$$

$$\bar{X} = \frac{14.96 + 15.05 + 15.04}{3} = 15.017$$

$$S = \sqrt{\frac{s_1^2 + s_2^2 + s_3^2}{3}}$$

$$S = \sqrt{\frac{1.52^2 + 1.49^2 + 1.53^2}{3}}$$

$$S = \sqrt{\frac{6.8714}{3}} = 1.513$$

Maximize Sample Size to Estimate the Population Average and Sigma

The greater the sample size n ("n is your friend"), the better your estimate of the population average and standard deviation!

TIP

Calculating Minimum Sample Size and Sensitivity, Variables Data

$$n = \left(\frac{Z \times S}{h}\right)^2$$ to calculate minimum sample size on variables data

n = minimum sample size on variables data (always round up)

Z = confidence level (When in doubt use $Z = 1.96$, per the following tip.)

S = the population standard deviation

h = the smallest change we want to be able to sense

(When in doubt, use h = total tolerance/10, or $h = 0.6S$.)

Note that the formula just given can be rewritten as:

$$h = \sqrt{\frac{Z^2 \times S^2}{n}}$$

This allows us to see what sensitivity h (change) we can expect to see with a given sample size and confidence level.

FORMULA

There is a lot of judgment that goes into calculating sample size. Often the final sample size will be a compromise between cost (both product and inspection) and customer need. The formulas allow you to make the decision on sample size knowledge-based rather than just a guess.

Looking at the components of the sample size formula, we see that the sample size n is influenced not only by the sensitivity (h) required, but also by the process sigma. Sometimes this formula will point out that the sample size requirement is so excessive that only a process improvement (reduced sigma) or a loosening of the customer's requirements (increased h) will make sampling viable. The alternatives to sampling include automatic inspection or 100 percent process sorting. This use of sample size formulas is a very productive use of the Six Sigma process.

Z, Confidence of Results

Z relates to the probability, or confidence, we are looking for. For one-tailed questions (like greater than), use $Z = 1.64$ for a 95 percent confidence level. On two-tailed problems (like greater than *or* less than), use $Z = 1.96$ for 95 percent confidence.

    We normally test to a 95 percent confidence level.

    Assume that a question is two-tailed unless it is specifically given as one-tailed. Therefore, *the Z will normally be 1.96.*

TIP

Here are some important notes with regard to confidence testing. First, you can find countless tables and computer programs that will test for significance at many different confidence levels. If you look at a low enough confidence level, you may find differences that are significant at that level that were *not* significant at a 95 percent confidence level. But at lower confidence levels, you are increasing the chance of erroneous conclusions. Also, using higher confidence levels will make little difference, since you usually have to rerun any test to see if the results hold. If you recall from the probability chapter, running two tests at a 0.95 probability reduces "chance" to $p = 0.05 \times 0.05 = 0.0025$! There is normally no reason to test to a higher confidence level than 95 percent, since you may miss real opportunity.

CASE STUDY: ENFORCING MINIMUM SAMPLE SIZE

A company had calculated the minimum sample size to be used for inspecting a skid of product. It assigned an inspector to randomly measure product as it was being put on the skid. The product measurements were entered into a computer, and the program then made a determination to ship the skid or to hold it for reinspection.

The product was produced 24 hours per day, 7 days per week. Supervision was minimal. Besides the four inspectors assigned to this line, there were relief inspectors and people who covered for vacations and illness.

Despite the incorporation of the minimum sample size instruction, the customer was still seeing skids that contained an excessive amount of defects. After several complaints, the supplier looked at past records of the quality checks and found that, at times, the quantity checked was far less than the calculated minimum sample size.

The company then changed the quality system to hold a skid when the sample size was too low, put a message on the inspector's computer screen that this was being done, and send an exception report to the quality manager. Instantly, the problem with low sample sizes stopped and the customer stopped seeing the defect excursions.

Sample Size Verification
Determining the minimum sample size for a quality system is not enough. Some system for verifying the sample size is also required.

TIP

Problem 3
Administrators of a high school had just gotten the results from a national achievement test, and they wanted to know how many random results they would have to review before deciding, with 95 percent confidence, whether the students' performance had *changed*. The historical sigma on this test was 1.24. They wanted to be able to sense a change of 0.6S, which is 0.744.

$$n = \left(\frac{Z \times S}{h}\right)^2$$
$$n = \left(\frac{1.96 \times 1.24}{0.744}\right)^2$$
$$n = 10.67$$

Answer: Therefore, they would have to look at the results from at least 11 tests to see, with 95 percent confidence, whether the performance had changed 0.744.

Problem 4
Administrators of a high school had just gotten the results from a national achievement test, and they wanted to know how many random results they would have to review before deciding, with 95 percent confidence, whether the students' performance had *improved*. The historical sigma on this test was 1.24. They wanted to be able to sense an improvement of 0.6S, which is 0.744.

Note that this problem is now one-tailed, because the administrators want to see only whether the students improved. This changes the value of Z to 1.64.

$$n = \left(\frac{Z \times S}{h}\right)^2$$

$$n = \left(\frac{1.64 \times 1.24}{0.744}\right)^2$$

$$n = 7.47$$

Answer: So, the administrators would have to look at the results from at least 8 tests to see, with 95 percent confidence, whether the performance had improved 0.744.

Note that by looking only for "improved" scores, the minimum sample size is reduced from 11 to 8. However, knowing only whether change occurred on the "up" side is usually not sufficient, since at some point the school will be concerned about change on both the "up" and the "down" sides. Because of this, the sample size is usually determined by a two-tailed $Z = 1.96$, as in Problem 3, giving a minimum sample size of 11.

Problem 5
Before doing the previous calculations, someone had already tabulated the results from 20 tests. At a 95 percent confidence level, what change in results (h) could be sensed from reviewing this many results versus the minimum 0.744 change target? The historical sigma on this test was 1.24.

$$h = \sqrt{\frac{Z^2 \times S^2}{n}}$$

$$h = \sqrt{\frac{1.96^2 \times 1.24^2}{20}}$$

$$h = 0.543$$

Answer: Therefore, the sensitivity on 20 results is 0.543 compared with the 0.744 target. The school officials would be able to sense a smaller change. This shows the benefit of a larger sample size, $n = 20$ versus $n = 11$.

USING A SAMPLE TO CHECK FOR A CHANGE VERSUS A POPULATION

We will use a three-step process:

1. Check the distributions to see whether the data histogram shapes (population versus sample) are substantially different.
2. If the distribution shapes are not substantially different, then see whether the sigma are significantly different.

3. If neither of these tests shows a difference, then check whether the averages are significantly different.

If we sense a significant difference at any point in these three steps, we need to stop and try to find the cause. Any significant difference is potentially important, since it can affect costs, quality, or some other factor.

Don't Use Data Analysis Alone to Drive Decisions   **TIP**

The following formulas will give you the ability to detect change. However, the reaction to any analytical finding should be tempered by common sense and expert knowledge. This should not keep you from pursuing a finding that violates common sense and expert knowledge, because it is not uncommon to discover that some preconceptions are invalid. However, tread softly, because the analysis could also be wrong.

Remember, we are testing to a 95 percent confidence level, which means that 5 percent of the time, you could be wrong!

---

1. Checking the Distributions

First, it may be necessary to plot a large number of individual measurements to verify that the distribution shape of the sample is similar to that of the earlier population. Although a process distribution will normally be similar over time, it is important to verify this, especially when running a test after a policy change, a machine wreck, a personnel change, or some other issue. We are concerned only about gross differences, such as one plot being very strongly skewed or bimodal compared to the other. If plotting is required, it will need a sample size of at least 36. If there is a substantial change in the distribution shape, there is no reason to do further tests, because we know that the process has changed and we should be trying to understand the cause and ramifications of the change.

If there are outliers (data values that are clearly separate from the general distribution), the causes of those data points must be determined before doing any quantitative statistics on the data. If you can confidently determine that the questionable data points are the result of an error in collecting or entering data and don't reflect the process, then the data points should be removed. If the wild data points are *not* an input error, then you have found a potential problem that must be resolved.

Examining Plotted Data   **TIP**

Visually examining plotted data will often give insights that can't be seen with any quantitative method.

2. Checking the Sigma

If the process distribution has not changed qualitatively (looking at the plots), then you can do some quantitative tests. The first thing to check is whether the sigma has changed significantly. The sigma on a process normally does not change unless a basic change in the process has occurred. To see if the sigma has changed, we calculate a *chi-squared test value* (Chi_t^2) (see Exhibit 12-1, page 139).

Chi-Squared Test Value of a Sample Sigma *s* Versus a Population Sigma *S*

$$Chi_t^2 = \frac{(n-1)s^2}{S^2}$$

n = sample size
s = sample sigma
S = population sigma

We compare the calculated Chi_t^2 results with the values in the simplified chi-squared distribution table (Exhibit 12-1). If the Chi_t^2 test value we calculated is less than the table low value or greater than the table high value, we are 95 percent confident that the sigma *s* of the sample is different from the sigma *S* of the population.

FORMULA

If there is a significant change in the sigma, there is no reason to do further tests; instead, we should be trying to understand the cause of the change and its ramifications.

3. Checking the Averages

If in steps 1 and 2 we did not find that either the distribution or the sigma had changed significantly, we are now free to test whether the sample average is significantly different from the population average. We will have to calculate a *t-test value* (t_t) to compare to a value in the simplified *t*-distribution table (Exhibit 12-2).

t Test of a Population Average $\bar{X}$ Versus a Sample Average $\bar{x}$

$$t_t = \frac{|\bar{x} - \bar{X}|}{\frac{s}{\sqrt{n}}}$$

$\bar{X}$ = population average
$\bar{x}$ = sample average
s = sample sigma
n = sample size

FORMULA

$|\bar{x} - \bar{X}|$ is the absolute value of the difference of the averages, so ignore a minus sign in the difference.

We then compare this calculated t-test (t_t) value against the value in the simplified t table (Exhibit 12-2). If our calculated t-test (t_t) value is greater than the value in the table, then we are 95 percent confident that the sample average is significantly different from the population average.

Exhibit 12-1. Simplified chi-squared distribution table to test a sample sigma s (with sample size n) versus a population sigma S

	95% Confident They Are Different if Chi_t^2 Is				95% Confident They Are Different if Chi_t^2 Is		
	< Low	Or	> High		< Low	Or	> High
n	Low Test		High Test	n	Low Test		High Test
6	0.831209		12.83249	36	20.56938		53.20331
7	1.237342		14.44935	37	21.33587		54.43726
8	1.689864		16.01277	38	22.10562		55.66798
9	2.179725		17.53454	39	22.87849		56.89549
10	2.700389		19.02278	40	23.65430		58.12005
11	3.246963		20.48320	41	24.43306		59.34168
12	3.815742		21.92002	42	25.21452		60.56055
13	4.403778		23.33666	43	25.99866		61.77672
14	5.008738		24.73558	44	26.78537		62.99031
15	5.628724		26.11893	45	27.57454		64.20141
16	6.262123		27.48836	46	28.36618		65.41013
17	6.907664		28.84532	47	29.16002		66.61647
18	7.564179		30.19098	48	29.95616		67.82064
19	8.230737		31.52641	49	30.75450		69.02257
20	8.906514		32.85234	50	31.55493		70.22236
21	9.590772		34.16958	55	35.58633		76.19206
22	10.28291		35.47886				
23	10.98233		36.78068	60	39.66185		82.11737
24	11.68853		38.07561	65	43.77594		88.00398
25	12.40115		39.36406				
26	13.11971		40.64650	70	47.92412		93.85648
27	13.84388		41.92314	80	56.30887		105.4727
28	14.57337		43.19452				
29	15.30785		44.46079	90	64.79339		116.989
30	16.04705		45.72228	100	73.3611		128.4219
31	16.79076		46.97922				
32	17.53872		48.23192				
33	18.29079		49.48044				
34	19.04666		50.72510				
35	19.80624		51.96602				

Exhibit 12-2. Simplified *t*-distribution table to compare a sample average (size = *n*) with a population average or to compare two samples of size n_1 and n_2, using $n = (n_1 + n_2 - 1)$. A 95 percent confidence level (assumes a two-tailed test) is used. If the calculated t_t test value exceeds the table *t* value, then the two averages being compared are different.

n	t value	n	t value	n	t value
6	2.571	26	2.060	45	2.015
7	2.447	27	2.056		
8	2.365	28	2.052	50	2.010
9	2.306	29	2.048		
10	2.262	30	2.045	60	2.001
				70	1.995
11	2.228	31	2.042		
12	2.201	32	2.040	80	1.990
13	2.179	33	2.037		
14	2.160	34	2.035	90	1.987
15	2.145	35	2.032		
				100+	1.984
16	2.131	36	2.030		
17	2.120	37	2.028		
18	2.110	38	2.026		
19	2.101	39	2.024		
20	2.093	40	2.023		
21	2.086				
22	2.080				
23	2.074				
24	2.069				
25	2.064				

Problem 6

We have made a process change on our infamous lathe that is machining shafts. We want to know, with 95 percent confidence, whether the "before" process, with an average $\overline{X}$ of 1.0003" and a sigma *S* of 0.00170", has changed.

Answer: We use our three-step process to look for change. Assume that we first plotted some data from after the change and compared them with a plot of "before" data and saw no large differences in the shape of the two distributions. We must now compare the sigma before and after the change.

What is the minimum sample size we need, assuming that we want to be able to see a change *h* of 0.6*S*?

$$h = 0.6S = 0.6 \times 0.00170" = 0.00102"$$

$$Z = 1.96$$

$$S = 0.00170"$$

$$n = \left(\frac{Z \times S}{h} \right)^2$$

$$n = \left(\frac{1.96 \times 0.00170"}{0.00102"} \right)^2$$

$$n = 11 \text{ (rounding up)}$$

Now that we know the minimum sample size, we can take the sample. We want to know whether the sample sigma is significantly different from the "before" population sigma. From the sample, we calculate $s = 0.00173"$.

$$n = 11$$

$$s = 0.00173"$$

$$S = 0.00170"$$

$$\text{Chi}_t^2 = \frac{(n-1)s^2}{S^2} = \frac{(11-1)0.00173"^2}{0.00170"^2} = 10.356$$

Looking at the simplified chi-square distribution table (Exhibit 12-1), with an $n = 11$, the low value is 3.24696 and the high value is 20.4832. Since our test value (10.356) is not outside that range, we can't say that the sigma of the sample is different (with 95 percent confidence) from the population sigma.

Since we were not able to see a difference in the distribution or the sigma, we will now see whether the averages are significantly different. Assume that the after-change sample ($n = 11$) had an average $\bar{x}$ of 0.9991".

$$\bar{x} = 0.9991"$$

$$\bar{X} = 1.0003"$$

$$s = 0.00173"$$

$$n = 11$$

$$t_t = \frac{|\bar{x} - \bar{X}|}{\dfrac{s}{\sqrt{n}}}$$

$$t_t = \frac{|0.9991" - 1.0003"|}{\dfrac{0.00173"}{\sqrt{11}}} = 2.3005$$

Looking at the simplified *t*-distribution table (Exhibit 12-2), with $n = 11$, our calculated *t*-test value (2.3005) is greater than the table value (opposite $n = 11$: 2.228). We therefore assume with 95 percent confidence that the sample average *is* significantly different from that of the population.

The average has changed significantly from what it was before. We should therefore decide whether the process change was detrimental and should be reversed. Here is where judgment must be used, but you have data to help.

CHECKING FOR A CHANGE BETWEEN TWO SAMPLES

We sometimes want to compare samples from two similar processes or from one process at different times. As in comparing a sample with a population, we do three steps in checking for a change between two samples.

1. Check the distributions to see whether they are substantially different.
2. If the distribution shapes are not substantially different, then see whether the sigma are significantly different.
3. If neither of these tests shows a difference, then check whether the averages are significantly different.

If we see a difference at any of these steps, it is important to know, since it affects costs, quality, or other such factors.

1. Checking the Distributions

First, it may be necessary to plot a large number of individual measurements to verify that the sample distribution shapes are similar. Although the distribution of a process will normally be similar over time, it is important to verify this, especially when running a test after a policy change, a machine wreck, a personnel change, or some other such event. We are concerned only about gross differences, such as one plot being very strongly skewed or bimodal relative to the other. If plotting is required, it will need sample sizes of at least 36. If there is a substantial change in the distributions, we know that the process has changed and that we should be trying to understand the cause and ramifications of the change.

2. Checking the Sigma

If the sample distributions have not changed qualitatively (looking at the plots), then you can do some quantitative tests. The first thing to check is whether the sigma has changed significantly. The sigma on a process does not normally change unless a substantial basic change in the process has occurred. To see if the sigma has changed, we do an *F* test.

F Test Comparing Two Samples' Sigma s

$$F_t = \frac{s_1^2}{s_2^2} \text{ (put the larger s on top, in the numerator)}$$

s_1 = sample with the larger sigma
s_2 = sample with the smaller sigma

The sample sizes n should be within 20 percent of each other. There are tables and programs that allow for greater differences, but since you can control sample sizes, and since you get more reliable results with similar sample sizes, these other tables and programs are generally not needed.

Compare this F_t with the value in the simplified F table (Exhibit 12-3). If the F_t value exceeds the table F value, then the sigma are significantly different.

Exhibit 12-3. Simplified F table (with 95 percent confidence) for comparing sigma from two samples (sizes = n_1 and n_2; sample sizes are equal within 20 percent).

$$n = \frac{n_1 + n_2}{2}$$

If the calculated F_t value exceeds the table value, assume that there is a difference.

n	F	n	F	n	F
6	5.05	26	1.96	60	1.54
7	4.28	27	1.93	70	1.49
8	3.79	28	1.90	80	1.45
9	3.44	29	1.88	100	1.39
10	3.18	30	1.86	120	1.35
11	2.98	31	1.84	150	1.31
12	2.82	32	1.82	200	1.26
13	2.69	33	1.80	300	1.21
14	2.58	34	1.79	400	1.18
15	2.48	35	1.77	500	1.16
16	2.40	36	1.76	750	1.13
17	2.33	37	1.74	1000	1.11
18	2.27	38	1.73	2000	1.08
19	2.22	39	1.72		
20	2.17	40	1.70		
21	2.12	42	1.68		
22	2.08	44	1.66		
23	2.05	46	1.64		
24	2.01	48	1.62		
25	1.98	50	1.61		

Problem 7

Suppose that in our now-familiar shaft example, we take two samples. (They could be from one lathe, or from two different lathes doing the same job.) Assume we plot the samples and find that the shapes of the distributions are not substantially different. We now want to know with 95 percent confidence whether the sample sigma are significantly different.

Sample 1	Sample 2
$\bar{x}_1 = 0.9982"$	$\bar{x}_2 = 1.0006"$
$s_1 = 0.00273"$	$s_2 = 0.00162"$
$n_1 = 21$	$n_2 = 19$

Answer: As before, we must first check the sigma to see whether the two processes are significantly different. We therefore calculate the F-test value and compare this with the value in the simplified F table (Exhibit 12-3).

Since our sample sizes are within 20 percent of each other, we can use the previous formula.

$$F_t = \frac{s_1^2}{s_2^2} = \frac{0.00273"^2}{0.00162"^2} = 2.840$$

We now compare 2.840 with the value in the simplified F table (Exhibit 12-3). Use the average $n = 20$ to find the table value, which is 2.17. Since our calculated value is greater than the table value, we can say with 95 percent confidence that the two processes' sigma are different. We must now decide what the cause and ramifications are of this change in the sigma.

Problem 8

Suppose that in our shaft example, we take two samples. (They could be from one lathe or from two different lathes doing the same job.)

Answer: We have already plotted samples and found that the distributions are not substantially different. We now want to know with 95 percent confidence whether the sample sigma are significantly different.

Sample 1	Sample 2
$\bar{x}_1 = 0.9982"$	$\bar{x}_2 = 1.0006"$
$s_1 = 0.00193"$	$s_2 = 0.00162"$
$n_1 = 21$	$n_2 = 19$

Calculating an F_t:

$$F_t = \frac{s_1^2}{s_2^2} = \frac{0.00193^2}{0.00162^2} = 1.42$$

We now compare 1.42 to the value in the simplified F table (Exhibit 12-3). Use the average $n = 20$ to find the table value, which is 2.17. Since 1.42 is less than the table value of 2.17, we can't say with 95 percent confidence that the processes are different (with regard to their sigma).

We now test to see if the two sample averages are significantly different.

3. Checking the Averages

Since we did *not* find that either the distribution shape or the sigma had changed, we now test whether the two sample averages are significantly different. We calculate a *t-test value* (t_t) to compare with a value in the simplified t distribution table (Exhibit 12-2).

t-Test Value of Two Sample Averages $\bar{x}_1$ and $\bar{x}_2$

$$t_t = \frac{\left|\bar{x}_1 - \bar{x}_2\right|}{\sqrt{\left(\dfrac{n_1 s_1^2 + n_2 s_2^2}{n_1 + n_2}\right)\left(\dfrac{1}{n_1} + \dfrac{1}{n_2}\right)}}$$

$\bar{x}_1$ and $\bar{x}_2$ are two sample averages.
s_1 and s_2 are the sigma on the two samples.
n_1 and n_2 are the two sample sizes.

$\left|\bar{x}_1 - \bar{x}_2\right|$ is the absolute difference between the averages, ignoring a minus sign in the difference.

We then compare this calculated t-test value against the value in the simplified t-distribution table (Exhibit 12-2). If our calculated t-test number is greater than the value in the table, then we are 95 percent confident that the sample averages are significantly different.

Returning to problem 8, we must calculate our test t_t:

$\bar{x}_1 = 0.9982"$ $\bar{x}_2 = 1.0006"$

$s_1 = 0.00193"$ $s_2 = 0.00162"$

$n_1 = 21$ $n_2 = 19$

$$t_t = \frac{\left|\bar{x}_1 - \bar{x}_2\right|}{\sqrt{\left(\dfrac{n_1 s_1^2 + n_2 s_2^2}{n_1 + n_2}\right)\left(\dfrac{1}{n_1} + \dfrac{1}{n_2}\right)}}$$

FORMULA

$$t_t = \frac{\left|0.9982" - 1.0006"\right|}{\sqrt{\left(\dfrac{21(0.00193")^2 + 19(0.00162")^2}{21+19}\right)\left(\dfrac{1}{21} + \dfrac{1}{19}\right)}} = 4.24$$

We now compare this 4.24 with the value from the simplified t-distribution table (Exhibit 12-2). (Use $n = n_1 + n_2 - 1 = 39$.) Since the calculated 4.24 is greater than the table value of 2.024, we can conclude with 95 percent confidence that the two process means are significantly different.

We would normally want to find out why and decide what we are going to do with this knowledge.

EXAMPLE SHOWING IMPORTANCE OF PLOTTING DATA

Here is an example showing why you can't just look at the numerical statistics, and why you first have to examine the plotted data. The following are two sets of 50 pieces of data, with the averages and standard deviations shown below each set of data. Notice that the averages and standard deviations of both groups of data are basically the same.

Data Set 1

0.9991	0.9991	0.9998	1.0009
0.9978	0.9993	1.0002	1.0009
0.9982	0.9993	1.0002	1.001
0.9984	0.9994	1.0003	1.001
0.9985	0.9995	1.0003	1.0012
0.9986	0.9996	1.0003	1.0013
0.9988	0.9996	1.0004	1.0014
0.9989	0.9996	1.0004	1.0015
0.9989	0.9996	1.0005	1.0015
0.999	0.9996	1.0006	1.0015
0.999	0.9998	1.0006	1.0016
0.9991	0.9998	1.0008	
0.9972	0.9998	1.0009	
Average		=	0.999892
Standard deviation		=	0.00105

Data Set 2

0.9992	0.9993	1.0000	0.9995
0.9992	0.9993	0.9995	1.0000
0.9992	0.9993	0.9995	1.0000
0.9992	0.9993	0.9996	1.0000
0.9993	0.9993	1.0000	1.0000
0.9993	0.9994	1.0000	1.0002
1.0000	0.9994	0.9993	1.0034
0.9993	0.9994	1.0000	1.0025
0.9993	0.9994	1.0000	1.003
0.9993	1.0000	0.9994	1.0032
0.9993	0.9994	1.0000	1.0021
0.9993	0.9994	1.0000	
0.9993	0.9995	1.0000	
Average		=	0.999896
Standard deviation		=	0.00105

Now we want to look at the data plots, shown in Exhibits 12-4 and 12-5, for both sets of data.

Exhibit 12-4. Histogram for data set 1

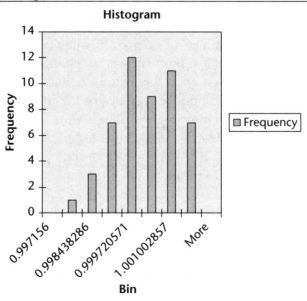

Exhibit 12-5. Histogram for data set 2

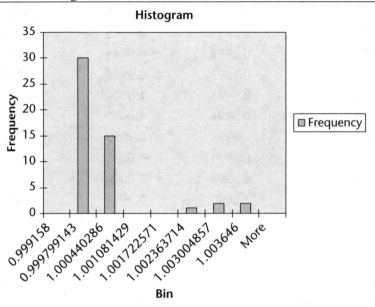

Note that the data distributions have completely different shapes. The distributions don't just have a wider spread, nor did their centers just change. They have totally different shapes.

These processes cannot be compared using standard numerical tests because they are two completely different processes. They might have been the same process at some time in the past—at least you thought they were, or else you probably wouldn't be comparing them now. However, because of a machine wreck or an unknown process change, they are no longer even similar. Trying to compare them now is like comparing apples to oranges. This difference cannot be seen by just looking at the data averages and standard deviations; plotting is required.

Note that discovering this difference is usually a good thing, because it often gives insight into some process deviation or change that was unknown and that should be addressed.

Tests on Averages and Sigma s Never Prove "Sameness"
The chi-squared, F, and t tests test only for significant difference. If these tests do not show a significant difference, it does not prove that the two samples or the sample and the population are identical. It just means that with the amount of data we have, we can't conclude with 95 percent confidence that they are different. Confidence tests never prove that two things are the same!

TIP

INCIDENTAL STATISTICS TERMINOLOGY NOT USED IN THE PREVIOUS TESTS

You will not find the term *null hypothesis* used in the previous discussion of confidence tests, but it is implied by the way the tests are done. *Null hypothesis* is a term that is often used in statistics books to mean that your base assumption is that nothing (null) changed. (An analogy is someone being assumed innocent until proven guilty.)

This assumption is included in the tests, and it is the basis for the tip declaring that the hypothesis tests never prove "sameness." (Again, just because a person is not proven guilty does not necessarily mean that he is innocent.) There is no need to add the complexity of the term *null hypothesis* when the nature of the tests implies it.

Several of the tables used in this book are called "simplified." This includes the chi-squared, F, and t tables. The main simplification relates to the column showing sample size n. In most other statistics books, the equivalent chi-squared, F, and t tables label this column as "degrees of freedom." One statistics book states that "degrees of freedom" is one of the most difficult terms in statistics to describe. The statistics book then goes on to show that, in almost all cases, "degrees of freedom" is equivalent to $n - 1$. This therefore becomes the knee-jerk translation ($n - 1$ = degrees of freedom) of almost everyone using tables with degrees of freedom.

The chi-squared, F, and t tables in this book are shown with the $n - 1$ equivalency built in. This was done to make life easier. In the extremely rare cases in which degrees of freedom is not equivalent to $n - 1$, the resultant error will be trivial in comparison with the accuracy requirements of the results. The validity of your Six Sigma test results will not be compromised.

You may need this ($n - 1$ = degrees of freedom) equivalency if you refer to other tables or use software with degrees of freedom requested.

Additional Practice Problems

Problem 9
An insurance company wants to see whether the costs for a certain medical procedure have changed in the last three months. The company has a great deal of historical data showing that the distribution shape has stayed the same over the years, even when costs changed. The company is not aware of anything that would have led to a change in the distribution. The historical average cost for this procedure was $387.61, with a sigma of $12.76. The company wants to be able to identify a cost change

as low as $4.00 and be 95 percent confident of the answer (so that it knows whether to change rates).

a. Will the company have to plot data to compare distributions?

b. What is the minimum number of random procedure costs that the company will have to sample? (Using the sample number n obtained earlier, the recent sample cost data on this procedure indicate an average cost of $391.04 and a sigma of $14.23.)

c. Does the sigma change indicate that the cost variation is no longer the same as the historical value?

d. If this test does not show a significant sigma change, is the average change significantly different?

Problem 10

After analyzing the previous sample, the insurance company decides to take a similar sample from a separate district to see whether the sample costs for the two districts are different. The two districts have a great deal of historical data showing that the distribution shapes are similar and have stayed the same over the years, even when costs changed. The company is not aware of anything that would have led to a change in the distributions. As given previously, the first district sample had an average price of $391.04 and a sigma of $14.23. The second district sample had an average price of $384.97 with a sigma of $16.06. Are the two districts different at 95 percent confidence?

Solutions to Additional Practice Problems

Problem 9

An insurance company wants to see whether the costs for a certain medical procedure have changed in the last three months. The company has a great deal of historical data showing that the distribution shape has stayed the same over the years, even when costs changed. It is not aware of anything that would have led to a change in the distribution. The historical average cost for this procedure was $387.61, with a sigma of $12.76. The company wants to be able to identify a cost change as low as $4.00 and be 95 percent confident of the answer (so that it knows whether to change rates).

a. Will the company have to plot data to compare distributions?

Technically no, because it has historical plots showing that the distributions are stable and no knowledge of anything that would have affected the distribution (such as a change in the procedure). However, it never hurts to do an additional plot, and, since the data are being collected anyway, it seems like a prudent thing to do.

b. **What is the minimum number of random procedure costs that the company will have to sample?**

$$n = \left(\frac{Z \times S}{h}\right)^2$$

$Z = 1.96$
$S = \$12.76$
$h = \$4.00$

$$n = \left(\frac{1.96 \times \$12.76}{\$4.00}\right)^2$$

$$n = 39.09$$

Thus, the insurance company will have to sample a minimum of 40 procedure costs.

Using the sample number n obtained previously, the recent sample cost data on this procedure indicate an average cost of $391.04 and a sigma of $14.23.

c. **Does the sigma change indicate that the cost variation is no longer the same as the historical value?**

$$Chi_t^2 = \frac{(n-1)s^2}{S^2}$$

$n = 40$
$s = \$14.23$
$S = \$12.76$

$$Chi_t^2 = \frac{(40-1)(\$14.23)^2}{(\$121.76)^2}$$

$$Chi_t^2 = 48.50$$

Looking at the simplified chi-squared distribution table (Exhibit 12-1), $n = 40$, the low-test value is 23.6543, and the high-test value is 58.12005. Since our calculated Chi_t^2 value of 48.50 is between these two numbers, we can't say with 95 percent confidence that the sigma has changed.

d. **If this test does not show a significant sigma change, is the average change significantly different?**

Since we did not find that the sigma has changed, we check the average.

$$t_t = \frac{|\bar{x} - \bar{X}|}{\frac{s}{\sqrt{n}}}$$

$$\bar{X} = \$387.61$$
$$\bar{x} = \$391.04$$
$$s = \$14.23$$
$$n = 40$$

$$t_t = \frac{|\$391.04 - \$387.61|}{\frac{\$14.23}{\sqrt{40}}}$$

$$t_t = 1.524$$

The simplified t table (Exhibit 12-2), $n = 40$, shows that the t value is 2.023. Since the calculated $t_t = 1.524$ is not greater than 2.023, we can't say with 95 percent confidence that the average has changed.

Problem 10

After analyzing the previous sample, the insurance company decides to take a similar sample from a separate district to see whether the sample costs for the two districts are different. The two districts have a great deal of historical data showing that the distribution shapes are similar and have stayed the same over the years, even when costs changed. The company is not aware of anything that would have led to a change in the distributions. As given previously, the first district sample had an average price of $391.04 and a sigma of $14.23. The second district sample had an average price of $384.97 with a sigma of $16.06. Are the two districts different at 95 percent confidence?

We do not have to plot the distributions, and, as calculated in the previous problem, the sample size for both samples is 40.

Checking the sigma:

$$F_t = \frac{s_1^2}{s_2^2} \text{ (put the larger } s \text{ on top, as the numerator)}$$

$$s_1 = \$16.06$$
$$s_2 = \$14.23$$

$$F_t = \frac{(\$16.06)^2}{(\$14.23)^2}$$

$$F_t = 1.274$$

The simplified F table (Exhibit 12-3), $n = 40$, shows that $F = 1.7$. Since our calculated F_t value of 1.274 does not exceed 1.7, we can't say with 95 percent confidence that the two samples' sigma are different.

We can now check to see if the two sample averages are significantly different.

$$t_t = \frac{|\bar{x}_1 - \bar{x}_2|}{\sqrt{\left(\dfrac{n_1 s_1^2 + n_2 s_2^2}{n_1 + n_2}\right)\left(\dfrac{1}{n_1} + \dfrac{1}{n_2}\right)}}$$

$|\bar{x}_1 - \bar{x}_2| = \$6.07$

$s_1 = \$16.06$

$s_2 = \$14.23$

$n_1 = 40$

$n_2 = 40$

$t_t = 1.79$

The simplified t table (Exhibit 12-2) shows that for $n = 80$ (the closest table value to $n_1 + n_2 - 1 = 79$), the t value is 1.99. Since the calculated $t_t = 1.79$ is not greater than 1.99, we can't say with 95 percent confidence that the two district averages are different.

Note that in this problem, another approach would have been to compare the second district sample with the first district population. The problem statement, however, specified comparing the two samples.

WHAT WE HAVE LEARNED IN CHAPTER 12

1. Valid sampling and analysis of variables data are needed for the Define, Measurement, Analyze, and Control steps in the DMAIC process.
2. Variables data are usually in decimal form, with no discrete "steps."
3. The rule-of-thumb minimum sample size for variables data is 11.
4. When possible, use the formulas to determine the minimum sample size.
5. We seldom have complete data on a population, but we use samples to estimate its composition.
6. The greater the sample size, the better the estimate for the population.
7. Sample size is usually a compromise between cost and desire for accuracy.
8. We normally work to a 95 percent confidence test level.

9. Assume two-tailed (high *and* low) problems unless a problem is specifically indicated as being one-tailed.

10. We normally want to sense a change equal to 10 percent of the tolerance, or $0.6S$.

11. When checking for a change, we can compare a sample with earlier population data or compare two samples with each other.

12. We follow a three-step process when analyzing for change. Compare distribution shapes first, then sigma, and then averages. If significant change is identified at any step in the process, we stop. We then must decide what to do with the knowledge that the process has changed.

13. Change analysis is useful on any process or population where data are available.

RELATED READING AND SOFTWARE

Lloyd R. Jaisingh, *Statistics for the Utterly Confused* (New York: McGraw-Hill, 2000).

T. D. V. Swinscow and M. J. Cambell, *Statistics at Square One*, 10th ed. (London: BMJ Books, 2001).

Mark J. Kiemele, Stephen R. Schmidt, and Ronald J. Berdine, *Basic Statistics: Tools for Continuous Improvement*, 4th ed. (Colorado Springs, CO: Air Academy Press, 1997).

MINITAB 13, Minitab Inc., State College, PA; www.minitab.com.

Testing for Statistically Significant Change Using Proportional Data

W hat you will learn in this chapter is how to use limited samples of proportional data. This chapter will parallel the last chapter, in which we learned how to use limited samples of variables data. Valid sampling and analysis of proportional data may be needed in all steps in the DMAIC process.

Changes in Real Processes, Proportional Data

Manufacturing. Use samples of scrap parts to calculate proportions on different shifts or similar production lines. Look for statistically significant differences in scrap rate.

Sales. Sample and compare proportions of successful sales by different salespeople.

Marketing. Use polls to prioritize where advertising dollars should be spent.

Accounting and software development. Use samples to compare the error rates of groups or individuals.

APPLICATIONS

Receivables. Sample overdue receivables, then compare proportions and due dates on different product lines. Adjust prices on products with statistically significant different overdue receivables.

Insurance. Sample challenged claims versus total claims in different groups, then compare proportions. Adjust group prices accordingly.

Proportional Data

As already defined in an earlier chapter, proportional data are based on attribute inputs, such as "good" or "bad," "yes" or "no," and other such distinctions. Examples are the proportion of defects in a process, the proportion of "yes" votes for a candidate, and the proportion of students failing a test.

DEFINITION

Data Type

Because of the large sample sizes required when using proportional data, use variables data instead whenever possible.

TIP

When people are interviewed on their preferences in an upcoming election, the outcome of the sampling is proportional data. The interviewer asks whether a person is intending to vote for a candidate, yes or no. After polling many people, the pollsters tabulate the proportion of yes (or no) results relative to the total number of people surveyed. This kind of data requires very large sample sizes. That is why pollsters state that, based on polling more than 1,000 people, the predictions are accurate within 3 percent or ±3 percent (with 95 percent confidence). We will be able to validate this with the following sample size formula (at 95 percent confidence).

Calculating Minimum Sample Size and Sensitivity, Proportional Data

FORMULA

$$n = \left(\frac{1.96 \sqrt{(p)(1-p)}}{h} \right)^2$$

n = sample size of attribute data, like "good" or "bad" (95 percent confidence)

p = probability of an event (the proportion of defects in a sample, chance of getting elected, or some similar classification) (When in doubt use $p = 0.5$, the most conservative.)

h = sensitivity, or accuracy required

(For example, for predicting elections, the sensitivity required may be ±3 percent, or $h = 0.03$. Another guideline is to be able to sense 10 percent of the tolerance or difference between the proportions.)

Note that the formula shown here can be rewritten as:

$$h = 1.96\sqrt{\frac{(p)(1-p)}{n}}$$

This allows us to see what sensitivity h we will be able to achieve with a given sample size and probability.

Let's do an election example. If the most recent polls show that a candidate has a 20 percent chance of getting elected, we may use $p = 0.2$. We will want an accuracy of ±3 percent of the total vote, so $h = 0.03$.

$$p = 0.2$$
$$h = 0.03$$
$$n = \left(\frac{1.96\sqrt{(p)(1-p)}}{h}\right)^2$$
$$n = \left(\frac{1.96\sqrt{(0.2)(1-0.2)}}{0.03}\right)^2$$
$$n = 682.95$$

So, we would have to poll 683 (round up) people to get an updated probability on someone whose estimated chance of being elected was 20 percent in earlier polls. However, if we had no polls that we could use to estimate a candidate's chances or if we wanted to be the most conservative, we would use $p = 0.5$.

$$p = 0.5$$
$$h = 0.03$$
$$n = \left(\frac{1.96\sqrt{(p)(1-p)}}{h}\right)^2$$
$$n = \left(\frac{1.96\sqrt{(0.5)(1-0.5)}}{0.03}\right)^2$$
$$n = 1{,}067.1$$

In this case, with $p = 0.5$, we would need to poll 1,068 (round up) people to be within 3 percent in estimating the chance of the candidate's being elected.

As you can see, the sample size of 683 with $p = 0.2$ is quite a bit less than the 1,068 required with $p = 0.5$. Since earlier polls may no longer be valid, most pollsters use the 1.068 as a standard. Using this formula, we have verified that pollsters require more than 1,000 inputs on a close election to be within 3 percent on forecasting the election outcome with 95 percent confidence.

Equally, because the forecast has only 95 percent confidence, the prediction can be wrong 5 percent of the time!

Here is an example showing how, in everyday life, conclusions are regularly made based on sample sizes that are too small. Calculate how many baseball games are needed, based on the teams' win/loss records, before you can state with 95 percent confidence that one team is 20 percent different from the others. At the start of the season, we will assume that all teams are equal, so we use the most conservative $p = 0.5$; h would then be 20 percent of p ($0.2 \times 0.5 = 0.1 = h$).

$$p = 0.5$$

$$h = 0.1$$

$$n = \left(\frac{1.96 \sqrt{(p)(1-p)}}{h}\right)^2$$

$$n = \left(\frac{1.96\sqrt{(0.5)(1-0.5)}}{0.10}\right)^2$$

$$n = 97$$

As you can see, the team needs to have played 97 games. Watch how early in the season fans talk about one team being far better or worse than the others and see how many of the calculated 97 games required have been played.

In some cases we have a choice of getting variables or attribute data.

Many companies choose to use go/no-go gauges for checking parts. They make this choice because go/no-go gauges are often easier to use than variables gauges that give measurement data. However, a go/no-go gauge checks only whether a part is within tolerance (that is, whether it is good or bad); it gives no indication as to how good or how bad the part is. This generates attribute data that are then used to calculate proportions.

Any process improvement is far more difficult with proportional data because it requires much larger sample sizes than the variables data used in the examples in the previous chapter would require. Using a gauge that gives variables (measurement) data output is a better choice!

With proportional data, comparing samples or a sample versus the population involves comparing proportions stated as decimals. These comparisons can be "defects per hundred" or any other criterion that is consistent with both the sample and the population. Although these proportions can be stated as decimals, the individual inputs are still attributes.

Using Proportional Data as Variables Data

If you have multiple periods (at least 11) and you have calculated proportions for each period (using the minimum sample size formulas), you can calculate an average and a sigma for these periods and treat the resulting values as you would treat variables data. This is valid because each of the 11 or more periods has decimal (continuous) values.

Comparing a Proportional Sample with the Population (95 Percent Confidence)

First, we must calculate a test value Z_t.

$$Z_t = \frac{|p - P|}{\sqrt{\dfrac{P(1-P)}{n}}}$$

P = proportion of defects (or whatever) in the population (historical)

p = proportion of defects (or same as above) in the sample

$|p - P|$ = absolute proportion difference (no minus sign in difference)

n = sample size

If $Z_t > 1.96$, then we can say with 95 percent confidence that the sample is different from the population.

The following is the formula for comparing two proportional data samples with each other. We will then show a case study that incorporates the formulas for calculating proportional data sample size, comparing a proportional sample with the population, and comparing two proportional samples with each other.

Comparing Two Proportional Data Samples (95 Percent Confidence)

Again, we must calculate a test value Z_t.

$$Z_t = \frac{\left|\dfrac{x_1}{n_1} - \dfrac{x_2}{n_2}\right|}{\sqrt{\left(\dfrac{x_1 + x_2}{n_1 + n_2}\right)\left(1 - \dfrac{x_1 + x_2}{n_1 + n_2}\right)\left(\dfrac{1}{n_1} + \dfrac{1}{n_2}\right)}}$$

x_1 = number of defects (or whatever) in sample 1

x_2 = number of defects (or same as above) in sample 2

$\left|\dfrac{x_1}{n_1} - \dfrac{x_2}{n_2}\right|$ = absolute proportion difference (no minus sign in difference)

n_1 = size of sample 1

n_2 = size of sample 2

If $Z_t > 1.96$, then we can say with 95 percent confidence that the two samples are significantly different.

CASE STUDY: TESTING WHETHER AN AUTOMATIC PACKER DAMAGED PRODUCT

A fragile component was automatically packed using a high-speed packing machine. The component was then shipped to another plant, where it was assembled into a final consumer product. This assembly took place in a complex machine that would jam if a component broke.

The assembly machine was experiencing excessive downtime, because 0.1 percent of the fragile components were breaking during assembly. The assembly plant suspected that the components were being cracked during the automatic packing process at the component plant. These cracks would then break during assembly, causing downtime.

In order to test this theory, a test was run in which half the product was packed manually while the other half was packed automatically. The product was assigned to the two packing processes randomly, making the populations of the automatically packed components and the manually packed components the same, other than any cracks resulting from the packing process.

The original goal was to use a minimum sample size that would be able to sense a 10 percent difference between the two test samples (manual and automatic packing). However, this sample was found to be extremely large, exceeding the realities of being able to run a good test.

As a compromise, it was decided to use a test sensitivity of 50 percent for the difference between the two samples. This approach was thought to be reasonable, since many people felt that the automatic packer was the dominant source of the cracks and that the difference between the two samples would therefore be dramatic. Even with the reduced sensitivity, the calculated minimum sample size was 15,352 manually packed components!

The tests were run, with 15,500 components being manually packed and 15,500 components being automatically packed. This was to be used to verify that the 0.1 percent breakage historical baseline had not changed.

The results were that 12 of the automatically packed components and 4 of the manually packed components broke when assembled. It was determined that this difference was statistically significant. Because of the costs required to rebuild the packer, the test was rerun, and the results were similar.

The automatic packer was rebuilt, and the problem of components breaking in the assembly operation was no longer an issue.

Since this case study incorporates all of the formulas we have covered concerning the use of samples on proportional data, we will use a series of problems to review in detail how the case study decisions were reached. Note that in the case study, the company not only checked the manual sample against the historical population defect level, but also collected a current automatically packed sample in order to verify that the defect level did not just happen to be lower (or higher) during the test.

Problem 1
Assuming that we wish to be able to sense a 10 percent defect difference between a proportional data sample of components and an earlier population (or another proportional sample), and the historical defect level is 0.1 percent, what is the minimum number of samples we must study? Assume 95 percent confidence.

$$p = 0.001$$
$$h = 0.0001 \text{ (which is 10\% of } p)$$
$$n = \left(\frac{1.96 \sqrt{(p)(1-p)}}{h} \right)^2$$
$$n = \left(\frac{1.96 \sqrt{(0.001)(1-0.001)}}{0.0001} \right)^2$$
$$n = 383{,}776$$

Answer: We would have to check 383,776 components to have 95 percent confidence that we could see a change of 0.01 percent.

You can see why the company concluded that this was an excessive number of components to test.

Problem 2

Assuming that we wish to be able to sense a 50 percent defect difference between a proportional data sample of components and an earlier population (or another proportional sample), and the historical defect level is 0.1 percent, what is the minimum number of samples we must study? Assume 95 percent confidence.

$$p = 0.001$$
$$h = 0.0005 \text{ (which is 50\% of } p)$$
$$n = \left(\frac{1.96 \sqrt{(p)(1-p)}}{h}\right)^2$$
$$n = \left(\frac{1.96\sqrt{(0.001)(1-0.001)}}{0.0005}\right)^2$$
$$n = 15{,}352 \text{ (round up)}$$

Answer: We would have to check 15,352 components to have 95 percent confidence that we could see a change of 0.05 percent.

Although this is still a high number to test, it is only 4 percent of the 383,776 we calculated were needed to see a change of 0.0001. The minimum sample size changes dramatically with a change in sensitivity h.

Problem 3

In the test, 12 of 15,500 (0.0774 percent) automatically packed components broke. This was less than the 0.1 percent that historically broke. Is the difference between this test sample with 12 defects and the historical 0.1 percent statistically significant?

$$P = 0.001$$
$$p = 0.000774$$
$$n = 15{,}500$$
$$Z_t = \frac{|p - P|}{\sqrt{\dfrac{P(1-P)}{n}}}$$
$$Z_t = \frac{|0.000774 - 0.001|}{\sqrt{\dfrac{0.001(1-0.001)}{15{,}500}}}$$
$$Z_t = 0.89$$

Answer: Since 0.89 is less than the test value of 1.96, we can't say with 95 percent confidence that the sample is different from the population. Therefore, the test sample with the 12 defects is not significantly different from the historical 0.1 percent defect rate.

Problem 4

In the test, 4 of 15,500 (0.0258 percent) manually packed components broke during assembly. Is this significantly different from the 0.1 percent that historically broke?

$$P = 0.001$$
$$p = 0.000258$$
$$n = 15,500$$

$$Z_t = \frac{|p - P|}{\sqrt{\dfrac{P(1-P)}{n}}}$$

$$Z_t = \frac{|0.000258 - 0.001|}{\sqrt{\dfrac{0.001(1 - 0.001)}{15,500}}}$$

$$Z_t = 2.92$$

Answer: Since 2.92 is greater than the test value of 1.96, we can say with 95 percent confidence that the sample is different from the population. Therefore, the manually packed test sample is significantly different from the historical 0.1 percent breakage rate.

Problem 5

A baseline sample was collected during the test; in this baseline sample, 12 of 15,500 automatically packed components broke during assembly. Of the manually packed components, 4 of 15,500 broke during assembly. Is the manually packed sample breakage statistically significantly different from the breakage in the baseline sample?

$$x_1 = 12$$
$$x_2 = 4$$
$$n_1 = 15,500$$
$$n_2 = 15,500$$

$$Z_t = \frac{\left|\dfrac{x_1}{n_1} - \dfrac{x_2}{n_2}\right|}{\sqrt{\left(\dfrac{x_1 + x_2}{n_1 + n_2}\right)\left(1 - \dfrac{x_1 + x_2}{n_1 + n_2}\right)\left(\dfrac{1}{n_1} + \dfrac{1}{n_2}\right)}}$$

$$Z_t = \frac{\left| \dfrac{12}{15,500} - \dfrac{4}{15,500} \right|}{\sqrt{\left(\dfrac{12+4}{15,500+15,500} \right) \left(1 - \dfrac{12+4}{15,500+15,500} \right) \left(\dfrac{1}{15,500} + \dfrac{1}{15,500} \right)}}$$

$$Z_t = 2.001$$

Answer: Since 2.001 is greater than the test value of 1.96, we can say with 95 percent confidence that the manually packed test sample is significantly different from the baseline sample.

The reason we checked the manually packed sample against both the historical population and the baseline sample is that, even though the baseline was not statistically different from the historical breakage rate, the baseline sample had a lower defect result that could have been different enough for the manually packed sample to pass the sample/population test but fail the sample/sample test. If this conflicting result had occurred, we would have had to make a decision whether or not to rerun the test.

As it was, because of the cost of rebuilding the packer, the test was rerun anyway, with similar results. The decision was then made to rebuild the packer.

Just for interest, after the packer was rebuilt, the test was run a third time. This time the automatic packer had zero defects, whereas the manually packed sample still had some defects. Apparently the manual packing was not as gentle as had been assumed, and it was causing some cracks.

CASE STUDY: ISOLATING BAD PRODUCT

A factory producing a glass component had a process that wasn't very robust. It produced, on the average, 15 percent bad product, and the outgoing quality relied on an inspector who was on-line to reject bad product before the components were packed by an automatic packing machine. This inspector sat in front of a conveyor belt, where the product went by in a single line at 50 pieces per minute, and looked at each piece without picking it up. If the inspector saw a defect, she picked up the product and discarded it.

The components were shipped to an assembly plant, where they were unpacked manually and loaded into an assembly machine. Since the components were unloaded manually, the assembly plant was able to do some incoming inspection, and it had agreed to accept product with a defect rate as high as 3 percent. The assembly plant's experience was that when the rate was higher than 3 percent, the incoming inspection was not capable of segregating all the defective components, so some got into the assembly machine. The defects would then cause a machine wreck or, even worse, cause the finished product to be defective.

The assembly plant had complained for many months that it was often seeing skids of components arrive with well over the 3 percent allowable defects. Since it didn't see the manufacturer taking any measurable corrective action, the plant began taking samples from the top layer of each incoming skid of glass components. It would inspect 250 pieces; if 15 or more defects were found, the plant would reject the whole skid, which had more than 3,000 components.

The manufacturer protested vehemently, saying that the sample was invalid, since it came from only the top layer of parts, and therefore, the manufacturer argued, was not representative of the whole skid. The customer didn't want to hear it! It was seeing fewer machine wrecks since it had started sampling incoming components and rejecting "bad" skids, so it was going to continue to do this. The assembly plant was rejecting 20 percent of the incoming product and, to add insult to injury, was subtracting the labor cost of this incoming sampling from its payment for the glass components. This cost was in addition to the costs for the resultant reinspection and for product losses due to the 20 percent returned product.

The management of the component plant asked its home-office engineering team to get involved and to find a more cost-effective solution to the ongoing conflict over component quality. Obviously the ideal solution was to fix the root cause, a nonrobust process. However, the plant had put a lot of time, hours, and money into this process over the years, and the engineering team, although confident, was not going to bet on being able to fix this process within three months, which was the time target it was given to resolve the problem.

The company was just beginning to implement the Six Sigma process, so the engineers decided to start using some of the Six Sigma tools. First, they defined the problem better. They had a meeting with representatives from both plants, including several operators and inspectors from the glass plant. After getting through some initial emotional outbursts, they were able to develop a realistic problem definition.

The first part of the problem definition was that, when the glass process was producing an unusually high level of defects, the on-line inspector was unable to segregate the defective components effectively. The second part of the problem definition was that, when defective components were returned several weeks after production, the operator did not know when these periods with higher defects had occurred, which made it more difficult to tweak the process to an acceptable defect level. The operator needed quicker feedback about when the quality was declining. Even though the operator looked at product regularly as it was being produced, his sample size was too small and the inspection was not rigorous enough to allow a valid judgment on defect level.

So, here was the project as a result of the project definition. The goal was to identify periods when the glass process was producing excessive defects and to notify the operator. The product produced during this period should be isolated and held for reinspection before being sent to the assembly plant. The problem was that there was no current measure of the process defect level that could be used to ascertain when it had reached an excessive level, since it had already been demonstrated that the on-line inspector was not effective in finding defects at high levels. What was needed was an additional quality sample of the product before inspection.

Of course, management had told the engineering team to solve this problem without additional plant labor. The team had some money for some minor on-line physical changes and computers, but the "no extra labor" directive was firm.

The engineers met with the plant on-line inspectors, and it was agreed that the inspectors could periodically pick up a random product, inspect it, and key the inspection result into the computer system. The rate agreed upon so as not to impede the current inspection efficiency was one product every 15 seconds. The thought was that the packed components would be held in a queue on a conveyor line coming out of the packing machine while the samples pulled from production were being gathered. Once the sample was large enough to make a statistically valid decision on its quality, then all of the packed components would either be released for shipment to the customer or be set aside for another inspection.

The team now had a means of collecting samples of the components as they were produced. However, what interested the customer was the quality of the product *after* it was inspected on-line. Therefore, the team had to figure some way to "predict" the outgoing quality (after the on-line inspection) based on the quality of the product coming to the inspector.

The team took samples of the product going to the inspector, immediately returning components to the production line after sampling them. They sampled the same product again after the on-line inspection. In this way, they were able to ascertain the effectiveness of the inspector on each type of defect. They found that the on-line inspection was approximately 90 percent effective in finding large defects, like chips, but only 30 percent effective on small defects, like small cracks in the rim of the product.

Since the samples pulled from the line for inspection every 15 seconds were going to be keyed into the computer in terms of defect type, this proportion of effectiveness was applied to each defect type to predict what level of defect would go on to the customer. For example, for every 10 large defects found by the inspector in the 15-second sample, it was assumed that 1 would get to the customer, and that 7 out of 10 small defects would get through to the customer. With the "effectiveness" correction applied to each defect, it was possible to predict the outgoing quality after the on-line inspection.

The customer had been inspecting incoming components in samples of 250 pieces, rejecting any skid that had 15 or more defects out of the 250. The sensitivity (h) on this sample size was 3 percent, so the actual defect level was between 3 percent and 9 percent, with 95 percent confidence. The glass plant decided to use the same sampling based on the predicted outgoing defect level. This wasn't very selective, but proportional data made any sampling difficult. The sample size could be increased, but then the hold time made the queue of held components too large for the conveyor coming out of the packing machine. As it was, at one sample every 15 seconds, it took 75 minutes to get the sample size of 250 components, with the result that up to 3,750 pieces were being held. (The actual time required to get the 250 samples was 62.5 minutes, but some allowance was made for missed samples.)

The customer had been rejecting 20 percent of the product. The engineering team had set a goal that not more than 10 percent of the product would be held for reinspection at the glass plant. This would save half the reinspection cost, plus the costs the customer was billing the plant for incoming sampling and shipping costs to return components labeled defective. The 10 percent reinspection goal was felt to be reasonable, because the new sampling plan included a provision to send the defect information to the operator. In the early meetings, the operators had indicated that the defect feedback would enable them to address process issues far sooner than they could without this information.

To ensure that the inspector would pick up a random product sample every 15 seconds, a photocell, a timer, and a small air cylinder were installed upstream of the inspector. Every 15 seconds, the next product that came in front of the photocell was displaced slightly by the air cylinder.

The inspector was to pick up this product as the sample to inspect. If the inspector found a defect, he keyed the defect type into the computer. To save the inspector unnecessary work, no keyboard entry was required when no defect was found. The inspector then set aside the sample, defective or good. To ensure that the inspector had picked up the sample, additional photocells downstream from the inspector verified that there was a space on the conveyor belt where the selected sample had been. If there was a space and the inspector had not entered a defect into the computer, then the inspected product was counted as "good." If there was no space, it was assumed that the inspector had not taken the selected sample. If, after 75 minutes of holding, there were not at least 250 samples that had been inspected, then the held components were put into reinspection because the sample size was insufficient. (An interesting note on this feature was that it was self-policing. If an operator's product was put into reinspection because the inspector had not taken sufficient samples, the operators were able to address the problem themselves. No action by supervisors was required.)

As this project was unfolding, there was quite a lot of negativity expressed at both plants. Many people at the supplier plant, including the quality manager and the production manager, felt that the system would just put all the products into reinspection. These people had no faith in the statistical procedures that were being applied and did not believe that the operator would do better with the simplified control chart information on defects. As stated earlier, the plant was just starting with Six Sigma, and they were not yet comfortable with the methodology. Meanwhile, many people at the customer plant also had little faith in this approach to the problem. They felt that the supplier should either go after the root cause (the nonrobust process) or add more inspection labor.

Once the project was implemented, everyone was anxious to see how it did. Well, it exceeded all expectations. The customer was so satisfied with the incoming product that it stopped sampling. The components being held for reinspection at the glass plant were 6 percent—less than the team's 10 percent goal.

The management of the glass plant was so happy with the project that, in the year following the project, it implemented a similar system on all similar production lines.

As was noted, when this project was implemented, Six Sigma was just being introduced. If this project were to be done now, more Six Sigma tools would be used. If a QFD and an FMEA had been used on this project, it probably would have reduced some of the negativity of the people, since more of them would have felt involved and their input would have been considered early in the project. Also, if they had been fully trained in Six Sigma, they would have had more faith in the power of the statistical methods that were applied.

Additional Practice Problems

Problem 6

A software company has a development group of 10 people whose primary job is to write code. However, on a rotating basis, the group members also answer user questions related to past programs. Since many of the code writers would prefer to be just writing code, the company wants to know whether the time spent answering these questions justifies assigning a person whose sole function would be to answer questions, not write code. Since the new position would involve taking a person away from the development group, management wants to make sure that the procedure would free up at least 10 percent (one person) of the time of the current group. Although the manager is not sure what percentage of time is being spent on answering questions, he is reasonably sure that it isn't more than 20 percent.

To get an estimate of the time spent on answering questions, the manager decides that, from time to time, at random, he will survey the 10 members of the development group to find out whether they are working on a customer question at that given time. So, each time he surveys, he gets a sample of 10 inputs. Assume that the total hours worked by the group in a day is 80 and that the program questions come in at a constant and uniform level.

How many times must the manager survey to get a group accuracy of 2 hours/day at a 95 percent confidence level? Assuming that he randomly surveys 10 times per day, how many days will it take him to get an answer?

Problem 7

Assume that in the survey in the previous problem, the manager finds that the group is spending 7 hours a day answering questions, so he decides not to reassign a person just to answer questions.

Six months later, an identical survey finds that the group is spending 9 hours a day answering questions. Is the result of this second survey statistically different from the result of the first survey?

What are the minimum results that the manager should see on the survey before he assigns a person to answer questions full time? Again, assume that the program questions come in at a constant and uniform level.

Solutions to Additional Practice Problems

Problem 6

A software company has a development group of 10 people whose primary job is to write code. However, on a rotating basis, the group members also answer user questions related to past programs. Since many of the code writers would prefer to be just writing code, the company wants to know whether the time spent answering these questions justifies assigning a person whose sole function would be to answer questions, not write code. Since the new position would involve taking a person away from the development group, management wants to make sure that the procedure would free up at least 10 percent (one person) of the time of the current group. Although the manager is not sure what percentage of time is being spent on answering questions, he is reasonably sure that it isn't more than 20 percent.

To get an estimate of the time spent on answering questions, the manager decides that, from time to time, at random, he will survey the 10 members of the development group to find out whether they are working on a customer question at that given time. So, each time he surveys, he gets a sample of 10 inputs. Assume that the total hours worked by the group in a day is 80 and that the program questions come in at a constant and uniform level.

How many times must the manager survey to get a group accuracy of 2 hours/day at a 95 percent confidence level? Assuming that he randomly surveys 10 times per day, how many days will it take him to get an answer?

First, realize that 2 hours/day is 2/80, or 0.025 of the group's day. Therefore, h will be 0.025.

$p = 0.2$. (Since the manager is sure that the time spent is not more than 20 percent, 0.2 is the most conservative p value.)

$$h = 0.025 \text{ (to get an accuracy of two hours/day)}$$

$$n = \left(\frac{1.96 \sqrt{(p)(1-p)}}{h} \right)^2$$

$$n = \left(\frac{1.96 \sqrt{(0.2)(1-0.2)}}{0.025} \right)^2$$

$$n = 984$$

Answer: Since the manager gets 10 replies every time he surveys, he must survey 99 times and get 990 replies. If he surveys 10 times per day, he will have his answer in slightly less than 10 days.

Problem 7

Assume that in the survey in the previous problem, the manager finds that the group is spending 7 hours a day answering questions, so he decides not to reassign a person just to answer questions.

Six months later, an identical survey finds that the group is spending 9 hours a day answering questions. Is the result of this second survey statistically different from the result of the first survey?

What are the minimum results that the manager should see on the survey before he assigns a person to answer questions full time? Again, assume that the program questions come in at a constant and uniform level.

We must calculate a test value Z_t.

First, we must identify the values for the inputs to the equation:

$n_1 = 990$ (the manager asks 99 times and gets 10 replies each time)

$n_2 = 990$ (the manager asks 99 times and gets 10 replies each time)

$x_1 = $ the number of times in the first survey that the developers were working on a customer problem

(Since 7 hours is 7/80, or 0.0875, of the group's daily hours, the developers must have replied that they were working on customer problems $0.0875 \times 990 = 87$ times.)

$x_1 = 87$

$x_2 = $ the number of times in the second survey that the developers were working on a customer problem

(Since 9 hours is 9/80, or 0.1125, of the group's daily hours, the developers must have been replied that they were working on customer problems $0.1125 \times 990 = 111$ times.)

$x_2 = 111$

$$Z_t = \frac{\left| \dfrac{x_1}{n_1} - \dfrac{x_2}{n_2} \right|}{\sqrt{\left(\dfrac{x_1 + x_2}{n_1 + n_2} \right)\left(1 - \dfrac{x_1 + x_2}{n_1 + n_2} \right)\left(\dfrac{1}{n_1} + \dfrac{1}{n_2} \right)}}$$

$$Z_t = \frac{\left| \dfrac{87}{990} - \dfrac{111}{990} \right|}{\sqrt{\left(\dfrac{87 + 111}{990 + 990} \right)\left(1 - \dfrac{87 + 111}{990 + 990} \right)\left(\dfrac{1}{990} + \dfrac{1}{990} \right)}}$$

$$Z_t = 1.798$$

Answer: Since 1.798 is not greater than the test value 1.96, we can't say with 95 percent confidence that the 9-hour survey result is significantly different from the 7-hour survey result.

As for assigning someone to just answer customer questions, the manager wants to be confident that the survey shows the need to be significantly greater than 8 hours a day. In this way, he would be sure that the dedicated position would be full-time.

To test this, set as the baseline that the population requires 8 hours a day from the group for answering questions and see what survey result is required if it is to be significantly different from that baseline. We will try a survey result of 10 hours for answering questions, since this is the 8-hour baseline plus the 2 hours we used for sensitivity h.

We must use the formula for comparing a sample with a population. We first determine the value of the inputs to the equation:

$p = 0.125$ (10/80 of the group hours, our assumption to test)
$P = 0.1$ (8/80 of the group hours, our baseline)
$n = 990$ (from the previous calculation)

$$Z_t = \frac{|p - P|}{\sqrt{\dfrac{P(1-P)}{n}}}$$

$$Z_t = \frac{|0.125 - 0.1|}{\sqrt{\dfrac{0.1(1-0.1)}{990}}}$$

$$Z_t = 2.62$$

Answer: Since 2.62 is greater than the 1.96 test value, we can say with 95 percent confidence that, with a survey answer of 10 hours, the loading is significantly different from the 8-hour baseline loading, so the assigned person would be loaded in terms of time.

When checking a sample versus a population, "P + sensitivity h" and "P − sensitivity h" will always show significance. This is because we set up the test to check for 95 percent confidence at the h sensitivity.

WHAT WE HAVE LEARNED IN CHAPTER 13

1. Valid sampling and analysis of proportional data may be needed during all the steps in the DMAIC process.
2. Proportional data are generated from attribute inputs such as yes/no and go/no-go.

3. Testing for statistically significant change with proportional data involves comparing proportions stated as decimals.
4. Proportional data require much larger sample sizes than variables data.
5. Sample size is usually a compromise between cost and desire for accuracy.
6. We normally work to a 95 percent confidence level.
7. We generally want to be able to sense a change of 10 percent of the difference between the proportions or 10 percent of the tolerance.
8. When checking for a change, we can compare a sample with earlier population data or compare two samples with each other.
9. Change analysis using proportional data is useful anywhere we have proportions but don't have variables data.
10. Proportional data can be compared the same way as variables data if we have a large number of periods with calculated proportions for each.

RELATED READING AND SOFTWARE

Lloyd R. Jaisingh, *Statistics for the Utterly Confused* (New York: McGraw-Hill, 2000).

T. D. V. Swinscow and M. J. Cambell, *Statistics at Square One*, 10th ed. (London: BMJ Books, 2001).

Mark J. Kiemele, Stephen R. Schmidt, and Ronald J. Berdine, *Basic Statistics: Tools for Continuous Improvement*, 4th ed. (Colorado Springs, CO: Air Academy Press, 1997).

MINITAB 13, Minitab Inc., State College, PA; www.minitab.com.

Testing for Statistically Significant Change Using Nonnormal Distributions

What we will learn in this chapter is that many distributions are nonnormal and that these distributions occur in many places, but that we can use the formulas and tables we have already reviewed to get meaningful information on any changes in the processes that generated these distributions. We often do Six Sigma work on nonnormal processes.

Nonnormal Distributions

Manufacturing. Any process that has a zero at one end of the data is likely to have a skewed distribution. An example would be data representing distortion of a product.

Sales. If your salespeople tend to be fall into two distinct groups, one experienced and the other inexperienced, the data distribution showing sales versus age is likely to be bimodal.

Marketing. The data showing dollars spent in different markets may be nonnormal because of a focus on specific markets.

Accounting and software development. Error-rate data may be strongly skewed based on the complexity or uniqueness of a program or accounting procedure.

Receivables. Delinquent receivables may be skewed based on the product or service.

Insurance. Costs at treatment centers in different cities may be nonnormally distributed because of varying labor rates.

In the real world, nonnormal distributions are commonplace. Exhibit 14-1 shows some examples. Note that these are plots of the individual parts measurements.

Exhibit 14-1. Examples of nonnormal distributions

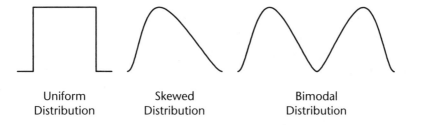

| Uniform | Skewed | Bimodal |
| Distribution | Distribution | Distribution |

Here are some examples of where these would occur:

- The *uniform distribution* would occur if you plotted the numbers occurring from spinning a roulette wheel or from rolling a single die.
- The *skewed distribution* would be typical of a one-sided process, like the nonflatness or positive warp on a machined part. Zero may be one end of the chart.
- The *bimodal distribution* can occur when there is play in a piece of equipment or when a process has a self-correcting feedback loop. This could also indicate that there are two independent processes involved.

In most classes on Six Sigma, one of the statistical tests is to check the plotted data to see whether they are "normal." However, it often seems that, no matter what the outcome of this test, the analysis of the data proceeds as if the data were completely normal. For most Six Sigma work, this is not a serious problem. Here is why.

If the population is not "normal," then the "absolute" probabilities generated from using computer programs or from tables may be somewhat in error. This would include results obtained from using the standardized normal distribution table (Exhibit 11-7). If we want to have good estimates of the absolute probabilities, the population must be normal if we are to use any computer program or table based on a normal distribution.

However, since most of the work we do in Six Sigma involves comparing similar processes relatively (before and after a particular event or between two similar processes) to see if there has been a significant change, these relative comparisons are valid even if the data we are using are nonnormal.

There are esoteric statistics based on nonnormal distributions and software packages that will give more accurate estimates of actual probabilities, but they require substantial knowledge of statistics. This added degree of accuracy is not required for most Six Sigma work, where we are looking for significant change, not absolute defect values.

Statistical Tests on Variables, Nonnormal Data   **TIP**

You can use the statistical tests included in this text to determine change, including referring to the numbers in the standardized normal distribution table (Exhibit 11-7) to compare a process before and after a change or to compare processes with similarly shaped nonnormal distributions. However, it is important to know that, although the relative qualitative comparison is valid, the absolute probability values for each process may be somewhat inaccurate.

As in checking processes with normal distributions, the distributions that are nonnormal must be plotted periodically to verify that the shapes of the distributions are still similar. If the shape of a distribution has changed dramatically, you can't use the formulas or charts in this book for before-and-after comparisons. It would be the proverbial apples vs. oranges thing! However, similar processes usually have and keep similarly shaped distributions.

CASE STUDY: TESTING FOR CHANGE ON NONNORMAL DATA

A production plant made a product that was not flat enough. The customer wanted product with a maximum 0.020″ nonflatness, a criterion that the production plant regularly exceeded. A plant engineer was assigned to solve this problem and came up with a new process that he felt would greatly reduce the nonflatness. The engineer had discovered that the nonflatness was directional, so he built a "reverse nonflatness" into the process to reduce the problem.

The production plant had nonflatness data (mean and sigma) from before the process change, and the engineer got additional measurements after the process change. He then used the before-and-after data to support a claimed 95 percent reduction in product that exceeded the 0.020″ maximum.

The customer questioned this conclusion, since the data were obviously skewed because zero was one end of the nonflatness data. To calculate his 95 percent reduction, the engineer had used tables based on a normal distribution; the customer took issue with that. The engineer defended his approach, since he was looking at differences between the before and after data on similarly shaped distributions. He was using the absolute defect rates from before and after only for calculating the relative reduction in defects between the two.

The customer then took the raw data that the engineer had used and redid the analysis using Crystal Ball 2000. This is a software package that enables users to fit the data to a skewed curve to get more accurate absolute defect levels. The percentage reduction that the customer calculated with the more difficult method was almost identical to the percentage reduction that the engineer had calculated.

As stated previously, the important thing to remember when comparing groups of data is that the two distributions can't have totally different shapes. It is critical that they have a normal distribution only when we want to use absolute values for things other than change comparisons.

Another important thing to know about nonnormal distributions is that you should understand the reason why the distribution is nonnormal. We have already mentioned some underlying causes, like having a zero at one end of the data. This would be the case when measuring the nonflatness of a plate. You would expect the data to be skewed, with many data points near zero and fewer points at the higher readings. If you get a nonnormal distribution when you don't expect it, it should set off an alarm for you to identify the cause.

WHAT WE HAVE LEARNED IN CHAPTER 14

1. We often do Six Sigma work on processes with nonnormal distributions.
2. You can use the statistical tests included in this book for calculating differences on similarly shaped nonnormal distributions.
3. The absolute probability values obtained may be somewhat inaccurate, but comparing probabilities to determine qualitative change in a process or in similar processes is valid.
4. As in checking processes with normal distributions, distributions that are nonnormal must be plotted periodically to verify that the shapes of the distributions are still similar. You cannot use standard statistical tests to compare differently shaped distributions.

5. Any distribution with a nonnormal shape should be analyzed to determine the cause of this shape. If the cause is not obvious, then an investigation should follow. Often large gains come from these surprise observations.

RELATED SOFTWARE

MINITAB 13, Minitab Inc., State College, PA; www.minitab.com.

Crystal Ball 2000, Decisioneering Inc., Denver, CO; www.decisioneering.com.

PART V

**Additional
Six Sigma Tools**

CHAPTER 15

Simplified Design of Experiments

What you will learn in this chapter is how to run a simplified design of experiments (DOE). The purpose of a DOE is to optimize a process by finding the right settings for a set of key process input variables (KPIVs). This chapter applies to the Improve step in the DMAIC process. It is primarily for those involved in manufacturing or process work.

Some Six Sigma practitioners feel that DOEs do not belong in a text for green belts, because DOEs are too complicated. Also, since there are whole books written just on this subject, these people don't believe that DOEs can properly be covered in a relatively few pages. These concerns may be valid for traditional DOEs, but the simplified DOE presented here has been used many times by green belts, with successful results.

First, some discussion on what a DOE entails. It is a controlled test of KPIVs, usually done right in the production environment using the actual production equipment. It attempts to measure all possible combinations of KPIVs, rather than taking a standard setup and modifying the variables one at a time, one after another. In this way, the DOE attempts to find any interaction among variables and includes this interaction in identifying the optimum settings. Many dedicated DOE software programs attempt to predict the optimum settings even if they are in between the actual test settings.

That sounds good, right? It is, but here are some of the challenges:

1. It is difficult to identify a limited list of KPIVs to test. After all, if the process were all that well understood, you would not have to run the DOE!
2. It is difficult to keep the variables that are *not* being tested under control. These could include temperature, humidity, operator skill, and other such factors.
3. A large number of test variables requires many trial iterations and set-ups. There are many reduced-iteration DOEs, but these all sacrifice statistical confidence.
4. One way in which a DOE can reduce the number of trials is to run with KPIV settings far outside the normal ranges. The problem with this approach is that many processes become nonlinear when this is done, and any conclusions become suspect. Some processes can't even be run outside their normal settings because of process limitations.
5. The results of the DOE must then be tested under controlled conditions, since the real test of a process change is its ability to predict future results.
6. What to use as an output goal is not trivial. What if product variation is reduced, but so is product output? What if the process settings give an excellent product, but require more operator skill? Most software programs limit optimization to one output measurement.

Now that we have listed all the reasons that a DOE may scare you, here is an effective way to run a simplified DOE that will minimize these difficulties and drive process improvement.

Note: The results of any DOE are usually *not* the key that drives the process improvement. Instead, it is the disciplined process of setting up and running the test that gives insight into the process. Observations made during the DOE often trigger process breakthroughs. *Serendipity* becomes dominant in this kind of test.

STEPS IN A SIMPLIFIED DOE

Here are the steps to run an effective simplified DOE.

1. Hold a meeting with representatives from every group that is familiar with the process. These people may include process operators, engineers, quality representatives, and even people from suppliers and customers. Have them develop the list of key process input variables (KPIVs) and prioritize them. Use the fishbone diagram, correlation check, and process flow diagram in this process. Pick no more than three variables to

test. If, after running the first simplified DOE, you feel that you may not have picked the critical three, have the same group reprioritize the variables and run another simplified DOE.

2. Test combinations of variables. To test two variables (A and B), each at two values (1 and 2), there are four possible combinations: A1/B1, A2/B1, A1/B2, and A2/B2. To test three variables, each at two values, there are eight combinations: A1/B1/C1, A2/B1/C1, A1/B2/C1, A1/B1/C2, A2/B2/C1, A2/B1/C2, A1/B2/C2, and A2/B2/C2.

3. What makes running a DOE difficult is that each of the combinations should be run a minimum of five times to get a valid statistical average for each iteration. This means that a test of two variables should have a minimum of $4 \times 5 = 20$ setups and a test of three variables should have a minimum of $8 \times 5 = 40$ setups. This assumes that the settings are run at the edge of, but within, the normal process window on each variable.

4. Do each setup independently of the earlier ones. It is not valid to count multiple readings in one setup as being the same as individual setups. Also, the setups should be in random order to reduce any influences of setup order.

Example of a Simplified DOE

Here is a simplified DOE to optimize the machining of a shaft so that the average diameter is as close as possible to 1.0000", which is critical to the customer. After each setup, the lathe will be reset at the nominal 1.0000" using a standardized setup gauge. Assume the historical sigma = 0.0010".

The first thing to do is to calculate the minimum number of shaft diameter readings that must be taken *at each setup*. This will determine the length of time each setup must be run. Using the equation from Chapter 12:

$$n = \left(\frac{Z \times S}{h} \right)^2 \text{ to calculate minimum sample size on variables data}$$

n = minimum sample size on variables data (always round up)
Z = confidence level (use $Z = 1.96$)
S = the population standard deviation (historically 0.0010")
h = the smallest change we want to be able to sense ($h = 0.6S = 0.0006"$)
$n = 11$

We know from past runs that if we run each setup for 0.5 hour, we will get at least 11 shafts during that 0.5 hour. So, for the simplified DOE, we will schedule the setup run length to be 0.5 hour.

Assume that, using a fishbone diagram, a knowledgeable group of people determined that machine tool design—rounded (R) or pointed (P)—and machining speed—fast (F) or slow (S)—are the two critical variables to test. The runs will each be 0.5 hour long. The test will be run seven hours per day for two days.

The four test combinations are R/F, R/S, P/F, and P/S. Each combination will be run 7 times on a random basis, which is more than the minimum of 5 required. This means there will be $4 \times 7 = 28$ setups. At each setup, we will measure 11 shafts, based on our calculations of minimum sample size. Therefore, we will have $28 \times 11 = 308$ individual shaft measurements being taken.

Exhibit 15-1 shows the results from the simplified DOE. The values in the table are the average and sigma of the 11 readings in each setup (in inches on the diameters). At the bottom of the table is the total of averages and sigma for the seven runs, calculated using the Chapter 12 formulas.

Exhibit 15-1. Results from example simplified DOE

Run #	R/F Avg	R/F s	R/F Avg	R/F s	R/F Avg	R/F s	R/F Avg	R/F s
1	1.0005	0.00087	1.0003	0.00021	1.0007	0.00091	1.0002	0.00022
2	0.9991	0.00102	1.0001	0.00034	0.9998	0.00099	0.9996	0.00032
3	0.9996	0.00090	0.9995	0.00031	0.9999	0.00082	0.9994	0.00037
4	0.9997	0.00094	1.0003	0.00028	0.9997	0.00097	0.9989	0.00042
5	0.9991	0.00076	1.0001	0.00027	1.0011	0.00081	1.0004	0.00027
6	1.0005	0.00111	1.0006	0.00032	1.0005	0.00101	0.9995	0.00039
7	0.9999	0.00089	0.9997	0.00023	1.0009	0.00088	1.0006	0.00025
Total Avg	0.99977		1.00009		1.00037		0.99980	
Total s		0.00093		0.00028		0.00092		0.00033

Since the goal of the simplified DOE is to minimize the difference from a nominal 1.0000" diameter, the results in Exhibit 15-1 are restated here to show how much they deviated (delta) from 1.0000":

Group	R/F	R/S	P/F	P/S
Delta average from 1.0000"	0.00023"	0.00009"	0.00037"	0.00020"
Sigma S on reading	0.00093"	0.00028"	0.00092"	0.00033"

We can see from the results that the R/S combination (rounded tool, slow machining speed) resulted in shafts closest to the nominal diameter, being off only 0.00009". This group also had the lowest sigma at 0.00028".

However, we must now check whether this group is significantly different from the next closest group, which is P/S.

We use the three-step procedure from Chapter 12 to test for a change between two samples.

The first step is to plot the data from the two groups to see whether the two distributions are substantially different. If they are dramatically different, we will know that we have two totally different processes, and we will not be able to use the following formulas to check for differences between the two groups. Then we would just use the raw data to make some judgment on which process is preferred or decide to run more tests at the two settings.

Since we ran each combination 7 times and had 11 measurements each time, we had 77 readings from each to plot, which is more than sufficient. Assume that these plots show that the distribution shapes are not substantially different.

We now do the second step, which is to compare the *sigma* of each group. Again, from Chapter 12:

F Test Comparing Two Samples' Sigma s

$$F_t = \frac{s_1^2}{s_2^2} \text{ (put the larger } s \text{ on top, in the numerator)}$$

$s_1 = 0.00033$
$s_2 = 0.00028$
$F_t = 1.389$

The sample sizes are both 77.

We now compare 1.389 with the value in the simplified F table (Exhibit 12-3). Use the average $n = 80$ (the closest value to 77 in the table) to find the table value, which is 1.45. Since 1.389 is less than the table value of 1.45, we can't say with 95 percent confidence that the processes are different (with regard to their sigma).

We should now see whether the R/S *average* is significantly different from the average of the next-best results (P/S). In Chapter 12, we learned the formula for comparing two sample averages:

t Test of Two Sample Averages $\bar{x}_1$ and $\bar{x}_2$

$$t_t = \frac{|\bar{x}_1 - \bar{x}_2|}{\sqrt{\left(\dfrac{n_1 s_1^2 + n_2 s_2^2}{n_1 + n_2}\right)\left(\dfrac{1}{n_1} + \dfrac{1}{n_2}\right)}}$$

$\bar{x}_1 = 0.00009"$
$\bar{x}_2 = 0.00020"$
$n_1 = 77$
$n_2 = 77$
$s_1 = 0.00028"$
$s_2 = 0.00033"$
$t_t = 2.230$

Compare $t_t = 2.230$ with the t value from the simplified t-distribution table (Exhibit 12-2, $n = 100+$, $t = 1.984$). Since $2.230 > 1.984$, we know with 95 percent confidence that the averages of the two groups are different.

Comparing the R/S group and the P/S group, we were not able to show that the sigma were different at a 95 percent confidence level. The significant difference was in the average. However, if the setup gauge were modified to account for the 0.00020" that the P/S group was off from the 1.0000" nominal, the P/S group might have performed as well as the R/S group. So, if there was any advantage in using the pointed tool (such as lower tool cost), then using the P/S setup should be considered.

Note that the fast speed (F) was worse with both tools. It would be advisable to run another test with the rounded tool with speed as the only variable, testing speeds that were slightly faster and slightly slower than the first test slow speed (S). Once the speed is bracketed, then it is necessary to retest at the final tool/speed combination to verify that the results replicate.

Some items of note. First, everyone involved should be present at the simplified DOE, observing and taking notes. For example, it may have been noted that the lathe vibrated at the higher speed. In that case, a lathe overhaul might allow the shaft to be machined at the higher speed with no loss of diameter consistency. This kind of observation during a DOE would not be unusual.

Pay Attention During a Simplified Design of Experiments
As many people as possible who have knowledge of the process should participate in a simplified DOE. They should be focused and should

TIP

not spend their time worrying about other problems. The reason this is important is that a DOE is a rare opportunity to observe a process under controlled conditions, and unexpected observations often trigger process breakthroughs. You need the best computers (brains) and the best sensors (eyes and ears).

Multiple simplified DOEs are often required, both to optimize results and to test the findings.

There are other considerations in reviewing the results of a simplified DOE. In the previous example, a slower machining speed caused less variation. But would the lower speed then raise costs? Maybe the most cost-effective option is to keep running at a higher speed, but to add some type of inspection device to reject exceptionally small or large shafts.

In the chapter on simplified QFDs, there was a case study that described designing a test in-line tubing cutter. Here is what happened during the DOE using this piece of test equipment.

Case Study: Simplified DOE on a Test Tubing Cutter

A simplified DOE was being run on a test in-line tubing cutter. The design of this machine was dictated by the output of a QFD that had been done some months earlier. One of the setups involved testing a support spring that had been added at the operators' insistence during the QFD. This involved a support wheel under the tubing that was to be supported with a spring.

The support spring that was initially picked to be run in the simplified DOE proved to be too weak to support the wheel. Someone had gone to get a stronger spring. Meanwhile, the manager went over to the wheel and supported it manually so that the process could run while they were waiting. This brought the usual jokes about the manager now being a critical part of the process, and the like.

The manager suggested that, rather than making jokes, the rest of the team should be paying attention and taking every process reading possible. Holding the wheel up with minimum force enabled the tubing to be cut perfectly. Having the minimum force supporting the wheel was critical. If the manager had not been observant, this would not have been noticed, and the KPIV probably would not have been discovered.

As you can see, observing the simplified DOE carefully and determining what to do with the results are as critical as using the correct procedure to set up the simplified DOE. The input from the operators during the QFD had caused the team to put the spring under the support wheel. However, the discovery that this spring had to be of minimum force came accidentally while the manager was supporting the wheel while someone was getting a stronger spring. The manager had played with the force as he was observing that he had a good cut, and he had gotten a sense that it was critical to use minimal force.

For those who choose to run a traditional DOE, there are many task-specific software programs available and many versions of DOEs. These include screening DOEs that help limit the number of input variables to run in a full DOE. Although these programs are powerful, they must be understood completely, and care must be taken when they are run. Screening DOEs sacrifice confidence, and reduced-iteration DOEs often test outside normal parameters. As stated previously, testing outside normal parameters can cause a process to be extremely nonlinear, making the predictions for optimum settings suspect. There is also a very real chance of damage to the equipment as a result of running at these unusual settings. There probably is a very good reason that the "normal" parameter limits are where they are; these reasons should be understood before ignoring them.

> **Use a Simplified DOE to Optimize a Machine or Process**
> If you have any machine or process that can be adjusted, the settings are probably not optimal unless someone has run a DOE. This is especially an opportunity to adjust those machines or processes for which there have been product complaints from customers.

TIP

WHAT WE HAVE LEARNED IN CHAPTER 15

1. This chapter applies to the Improve step in the DMAIC process.
2. Simplified DOEs can give valuable process knowledge.
3. The benefit from any DOE often comes from the discipline of running a controlled test, rather than from the direct output of the DOE.
4. It is important to limit the test variables to three or fewer. Use a fishbone diagram to assist in identifying the variables to be tested.
5. The output goal to be optimized is limited to one.

6. It may require two runs of the simplified DOE to optimize and verify the results.
7. Everyone involved in a simplified DOE should pay close attention to detail, looking for any surprises.
8. DOEs with abbreviated setups or runs always sacrifice statistical confidence in the results.
9. The formulas covered in previous chapters are sufficient to run a simplified DOE. There are software packages that are specifically designed for traditional DOEs; however, the user *must* take the time to understand all their quirks!

RELATED READING

Mark J. Kiemele, Stephen R. Schmidt, and Ronald J. Berdine, *Basic Statistics: Tools for Continuous Improvement*, 4th ed. (Colorado Springs, CO: Air Academy Press, 1997).

Douglas C. Montgomery, *Design and Analysis of Experiments*, 5th ed. (New York: John Wiley, 2001).

Stephen R. Schmidt and Robert G. Launsby, *Understanding Industrial Designed Experiments* (with CD-ROM), 4th ed. (Colorado Springs, CO: Air Academy Press, 1997).

CHAPTER 16

Simplified Control Charts

What you will learn in this chapter is how to make simplified control charts that are intuitive and that improve process stability. This tool is used in the Improve and Control steps of the DMAIC process. It is primarily for those involved in manufacturing or process work.

Control charts have had a chaotic history. In the 1960s, when Japan was showing the United States what quality really meant, some people here tried to implement some quick fixes, which included control charts. Few of them understood control charts, and control charts were never used enough to realize their full potential.

Control Chart
A control chart is a tool for monitoring variance in a process over time. A traditional control chart is a chart with upper and lower control limits on which the values of some statistical measure for a series of samples or subgroups are plotted. A traditional control chart uses both an average chart and a sigma chart.

DEFINITION

Now, with Six Sigma, control charts are getting a second look and have had some impressive successes. However, in a society that is geared to making product within tolerances and with a workforce that generally thinks that "sigma" is a health insurance company, control charts are a hard sell. But their potential has been proven, and there is a way to implement them that people will more readily accept and use. That way is the simplified control chart.

Traditional control charts use two graphs. The simplified control chart has one graph. The rationale for this simplification is covered throughout the chapter.

In most cases, if a supplier can reduce defect excursions (incidents with higher-than-normal quality issues), the customers will be happy with the product. This does not necessarily mean that all products will be within specification. It means that the customers have designed their processes so that they work with the normal incoming product. Therefore, the emphasis should be on reducing defect excursions, which is the reason control charts were designed.

TRADITIONAL CONTROL CHARTS

Traditional control charts have two graphs.

The top graph shows the process average, with statistically based control rules and limits telling the operator when the process is in or out of control. This graph cannot have any reference to the product tolerance because it displays product averages. By definition, a tolerance is for individual parts and is meaningless on averages. When tolerances are displayed on traditional control charts, which people sometimes try to do, they make no sense and can give an erroneous message that a process is running well when it isn't.

> **Displaying Tolerances on Control Charts** **TIP**
> A tolerance, or specification limit, can't be displayed on a chart that is displaying averages rather than the individual data values. It is not valid. Specifications refer to measurements of individual parts, not averages.

The second graph on traditional control charts is based on the process variation (or sigma), also with rules and limits.

If either of these graphs shows an out-of-control situation (based on a myriad of statistical rules), the operator is supposed to work on the process. However, the charts are somewhat confusing to the operators. Often the control chart shows an out-of-control situation even though quality checks do not show product that is out of specification. Also, one graph can show that the process is in control while the other is showing that it is out of control.

SIMPLIFIED CONTROL CHARTS

A single chart, such as the simplified control chart, can give the operator process feedback in a format that's understandable and intuitive and that encourages him to react before the product is out of specification. It is intuitive because it shows the average and the predicted data spread on one bar. This is the way an operator actually thinks of his process.

Here is how it works.

Assume that the operator or an inspector is regularly getting variables data on a critical dimension. Without regular variables data that are entered into some kind of computer or network, control charts are not effective.

Sometimes people use control charts with proportional data based on attributes (for example, defects/total), but if you look at the formulas in this book, you will see that when this is done, the error is generally huge because of the large sample size required with proportional data to get a statistically valid view of the population.

Setting the control limits on a simplified control chart should not be done in a vacuum. Data on the process should be gathered for a period of time before the control chart is implemented. Identify a time period when the process is producing acceptable product for an extended period and the process is reasonably stable. You will want a period that has at least 100 data points. Use the data from that stable time period to calculate the initial control limits. Remember, these are only the initial control limits. They will be reviewed later.

The initial control limits should not be tighter than needed to eliminate excursions. The intent is to run the current process the best it reasonably can, not to put undue (or unfair) stress on the operator. Exhibit 16-1 shows an example of how the initial control limits should be established.

Calculating Initial Control Limits   TIP

You should use at least 100 individual data points from an apparently in-control time period to calculate the initial control limits. Identify a time when the process is producing acceptable product for an extended period and the process appears stable, as in Exhibit 16-1.

Calculate the average $\overline{X}$ and sigma S of the 100-plus data points in the chosen data group.

We now have to assess whether the process is capable of running product within the specifications. Subtract the average $\overline{X}$ you just calculated

Exhibit 16-1. Example for establishing initial control limits

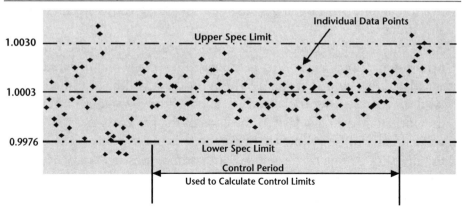

from the nearest specification limit. Divide the result by the sigma S you just calculated to get the number of sigma that "fit." This number indicates how well the process runs when it is in control.

Ideally this "fit" number will be greater than 3. A fit of ±3 sigma would give 99.7 percent product within specification, in accordance with the standardized normal distribution table, Exhibit 11-7. A fit of 2.5 sigma is 99 percent within tolerance. A fit of 2.0 sigma is 95 percent within specification. Without a process change, this is the best you can do! We will first assume that the fit is better than 3 sigma.

We will now set up the control chart (Exhibit 16-2). The initial upper control limit will be midway between the upper spec limit (upper tolerance) and $(\overline{X} + 3S)$. The initial lower control limit will be midway between the lower spec limit and $(\overline{X} - 3S)$. This will give the operator some "early

Exhibit 16-2. Simplified control chart

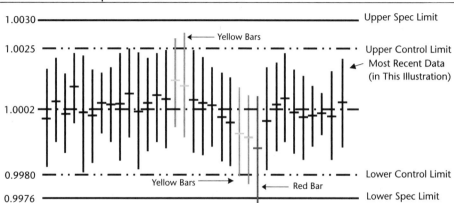

warning" that she is getting close to running product that is out of specification, while minimizing "false alarms" that cause needless process adjustments.

Each vertical feedback bar to the operator will be based on the data from the previous 11 product readings. The average $\bar{x}$ and sigma s will be calculated for these 11 readings. The vertical bar will have a small horizontal dash representing the average $\bar{x}$ and will be ±3 sigma in height from the horizontal average dash. If the ends of the bar cross a *control* limit, the vertical bar will be yellow (light gray in this book). This is the trigger for the operator to review the process and adjust it as necessary. If a vertical bar crosses a *spec* limit, it will be red (dark gray in this book). This indicates that some of the product is predicted to be out of specification.

Setting up this graph will require some computer skills. Exhibit 16-2 gives an example of this type of simplified control chart.

For reference, Exhibit 16-2 assumed an $\bar{X} = 1.0002$ and an $S = 0.0006$. This determined that the upper control limit was 1.0025, which is midway between the +3-sigma value of 1.0020 and the upper spec limit of 1.0030. The lower control limit of 0.9980 is midway between the −3-sigma value of 0.9984 and the lower spec limit of 0.9976.

A tolerance limit *can* be shown on this chart because each vertical line represents the projected ±3-sigma measurement spread for an individual part. This is equivalent to displaying individual data points. However, this chart also displays a control limit, to which the operator has to learn to respond. Note that the control limits must be *inside* the tolerance limits or you are asking the operator to do the impossible with the process! If the control limits are *outside* the tolerance limits, the target to which you are asking the operator to run is beyond the current consistent capability of the process.

Every time a simplified control chart is implemented, many people will protest that the operators must be given specific process instructions on how to bring the process back into control if it goes out. That's emphatically not true! The operators know how to do this. It may take them some time, but they will do it. They already react when the parts get out of specification; you just want them to respond somewhat sooner. This quicker reaction often saves them work, because it will require only tweaks rather than major process changes.

You will recall that we said that the initial data points were used to set the initial control limits. Once the process is actually running with the feedback benefits of control charts, the control limits should be reviewed.

Often they can be tightened because the process is running so much better. This will give operators an earlier warning when the process is beginning to drift. In rare situations, however, the control limits must be loosened because the process is always "in yellow" (in this graph, however, in light gray).

An operator using this kind of control chart will quickly learn that he can see trends in either the average or the variation and use that information to help debug the process. The information is intuitive and requires little training. This is especially critical when personnel changes are frequent. This is a key advantage of the simplified control chart over the traditional control chart, which is far less intuitive.

When we set up the control limits, we assumed that the process specification was greater than the $\overline{X} \pm 3S$ that we calculated from the in-control process. If the specification had *not* been greater, the control chart could still have been used, but we would have had to adjust everything for a reduced quality level. For example, if the fit had allowed only 2.8 sigma between the $\overline{X}$ and the nearest tolerance limit, the upper control limit could have been set midway between the upper spec and $(\overline{X} + 2.5S)$. The lower control limit would then have been midway between the lower spec and $(\overline{X} - 2.5S)$. Also, the vertical display bars that were calculated based on the last 11 readings would have been calculated at ± 2.5 sigma. However, 3-sigma limits are the norm, and that should be your target.

Case Study: A Simplified Control Chart's Perceived Six Sigma

A customer was unhappy with a supplier's product, deeming it only a 2.5-sigma process (99 percent within tolerance). In response to the customer's complaints, the supplier's quality manager began measuring product and identifying time periods when the process was running well and was seemingly in control.

When the quality manager taking these measurements suggested implementing simplified control charts, the response from the engineers, operators, and plant manager was that the plant could not use control charts without some specific process directions. They wanted the quality manager to give them specific process directions before implementing what they all thought would be a failure. However, the quality manager was insistent, and they all finally agreed to try simplified control charts.

The feedback system and control chart display screen were designed, and the quality manager met with the operators to get them "on board."

(This product was run 24 hours a day, 7 days a week, so four operators were involved.) The operators were largely noncommittal during these meetings.

During the first trial, only one of the operators was able to run within the control limits. The quality manager then met with each operator individually, explaining that one operator had been able to use the simplified control charts successfully. He told them that he wanted to do another trial and asked for their cooperation in trying to use the simplified control charts. On the next trial, two operators ran in control. By the third trial, all four operators were running in control. The system was then made permanent.

After six months, the results were analyzed. First, to the surprise of all, yields had actually improved with the simplified control charts. This was apparently because the tweaks that the operators now made to the process were less disruptive than the major changes that the operators had been making. Also, their customer now deemed the company a 6-sigma supplier (three defects per million product). This really surprised the supplier, because the in-house quality checks indicated that it was supplying a 3.5-sigma product (465 defects per million). Apparently the specification was tighter than what the customer required, so just eliminating the out-of-control defect spikes satisfied the customer.

Use Simplified Control Charts on Equipment or Process Output

TIP

Any piece of equipment or process that makes product that is generally acceptable to the customer, with the exception of defect excursions, is a natural for simplified control charts.

Although the most common benefit derived from the simplified control charts is reduced defect excursions, its inherent feedback feature often helps drive process insights and breakthroughs. The following case study was the first time my team used the simplified control chart. This project shows that even operators who are assigned to a line periodically can easily understand the simplified control chart.

CASE STUDY: CONTROL CHART VANQUISHES LOSSES

A component manufacturing plant had an ongoing issue with a high-volume product. The problem was warping of a surface that was critical to the customer. The customer had rejected 1,000,000 of these products in the previous year because of this defect. The component manufacturing plant had not been successful in keeping the warp within specifications, so it started a project to build an in-line inspection device to reject all products that exceeded the warp specification.

Although the plant production people recognized the need for this inspection device, they were afraid that it would reject more than 1,000,000 products per year because they had a strong suspicion that the customer was unknowingly using a lot of warped product that was out of specification. The plant would then be losing product in addition to what the customer was already rejecting.

To address this concern, the project to set up the automatic inspection included a plan to use the output data from the inspection device to give the operator feedback on warping. The plan was to display the warp severity on a computer monitor at the operator's station. This display was to be in the form of a simplified control chart, in the hope that this would help the operator reduce the degree of warp.

The simplified control chart proved to be more valuable than anyone had predicted. Apparently one operator knew the "secret" to reducing warp in the process. He was not a regular operator on this line, so his success in minimizing warp had not been noticed. Either no one had believed him or he hadn't gone out of his way to tell others. However, when he was an operator on this line after the simplified control chart was in place, his success was obvious. Soon others began to learn and use his process "secret."

The on-line automatic inspection combined with the improved process drove returns from the customer to zero within a few months. Also, the simplified control charts combined with the new process knowledge caused the warp losses from the automatic inspection device in the component plant to be very low.

CONTROL CHARTS FOR KPIVs

If the knowledge of a process is such that the key process input variables (KPIVs) are known, then simplified control charts can be used on the input variables rather than on the output. If the KPIVs are known, then the specification on each KPIV would be established and a simplified control chart would be implemented for each input variable just as shown previously for the output. This is a preferred way to control a process, but such detailed knowledge of the KPIVs is rare.

If Uncomfortable, Go Traditional
Anyone who is uncomfortable with the simplified control chart because it's unconventional should use a traditional control chart. At the end of this chapter are two excellent references for traditional control charts. My experience is that the simplified version has many advantages, but *any* control chart will provide many of the benefits described in this chapter. It is one of the few tools that almost *always* generates improvement.

TIP

WHAT WE HAVE LEARNED IN CHAPTER 16

1. This chapter is used in the Improve and Control steps of the DMAIC process.
2. Traditional control charts that use both an average chart and a sigma chart are somewhat confusing to operators. Since only relatively few are in use, the full potential of control charts has not been realized.
3. Operators want to see specification limits on a control chart, which are not valid on a chart that reflects averages rather than individual readings.
4. The simplified control chart shown in this chapter is intuitive to an operator and satisfies the previous two concerns.
5. Variables data and a computer system are needed for simplified control charts.
6. Control charts work best on processes that make products that are generally acceptable, other than for defect excursions. However, the feedback feature of control charts also drives process insight into any process.
7. Process instructions on how to react to an out-of-control situation are not required to implement a simplified control chart.
8. Some computer skills are required to set up the simplified control chart screen, data feedback, and other features.

RELATED READING

Hitoshi Kume, *Statistical Methods for Quality Improvement* (New York: Chapman & Hall, 1995).

Robert T. Amsden, Howard E. Butler, and Davida M. Amsden, *SPC Simplified: Practical Steps to Quality*, 2nd ed. (New York: Quality Resources, 1998).

PART VI

Statistical Tools for Design for Six Sigma

CHAPTER 17

What Tolerance Is Really Required?

What you will learn in this chapter is that most tolerances have *not* been determined based on the needs of the application. Tolerances often have little to do with what is really required. Therefore, tolerances can and should be questioned. You will also learn the RSS (root sum-of-squares) approach to calculating stacked tolerances. Tolerances apply to the Improve and Control steps in DMAIC. This chapter is primarily for those involved in design, manufacturing, or process work.

The initial tolerance on a part is often based on the machine that the designer believes will be used to make the part. The tightest tolerance that the machine can achieve is then used. Or, the designer just copies the tolerance from another similar part. An analysis of the real need is seldom done.

Sometime later, the customer may have some issue with the part and complain. When this happens, the tolerance is tightened—whether or not the issue actually involved the tolerance. Because most tolerances have evolved in this manner, there is large savings potential in determining what tolerance is really required.

CASE STUDY: REQUIRED TOLERANCES ON MOLDS

Eight sets of interchangeable molds were used on several high-speed machines. These mold sets were very expensive ($100,000 per set), so they tended to be kept for many years. However, the customer complained that the "old" mold sets were making product with too much variation and wanted the old mold sets replaced.

When an attempt was made to correlate product variation with specific mold sets, however, it was found that the age of a mold set had no correlation with product variation. Instead, it was concluded that several mold sets had been made with excessive variation from the start. Confidence tests, as covered in Chapter 12, confirmed this.

The mold shape was very complex, and the plant purchasing the molds was not capable of validating the mold dimensions. It had to use an outside firm with three-dimensional measuring capability to check the dimensions. When molds from each set were measured, it was found that *all* the sets had some molds that were outside specifications. The problem mold sets just had more molds that were further out of tolerance than the other sets!

Over the years, in response to complaints, the mold tolerances had been continually tightened until the supplier was not capable of making the molds within specifications. However, since the plant kept buying molds, the supplier kept selling them. The mold tolerances became meaningless.

Molds that were making acceptable product were studied, and it was found that the supplier was capable of making molds that made acceptable product, even though many of these molds were outside the specified tolerances. The tolerances were then doubled to reflect the dimensions of the molds that made acceptable product, based on this real-need criterion.

Inspection procedures were put in place to verify that all future molds were within the new specifications. Molds that were beyond the revised tolerances were taken out of production, and excess product variation was eliminated. With the more realistic mold specifications, the plant was able to find alternative suppliers, and the mold costs were reduced 50 percent. Total savings were in excess of $100,000 per year.

TOLERANCE STACK-UP

When multiple parts are "stacked" and have a cumulative tolerance buildup, the traditional way to handle the stack-up variation is to assume the "worst case" for each component. (Allow for all parts being at the high end of the tolerance or all parts being at the low end of the tolerance.) "Stacked" parts are akin to having multiple blocks, placed one block on top of another.

> ## Tolerance Stack-Up Analysis
> This is the process of evaluating the effect that the dimensions of all components can have on an assembly. There are various methods used, including worst-case, RSS (root sum-of-squares), modified RSS, and Monte Carlo simulations.
>
> DEFINITION

Example, Part A

Assume that there is a stack of 10 parts, each a nominal 1.000" thick, and that the tolerance on each part is ±0.010" (the total tolerance on each part = 0.020").

A traditional worst-case design using this stack of parts would assume:

$$\text{Maximum stack height} = 1.010" \times 10 = 10.100"$$

$$\text{Minimum stack height} = 0.990" \times 10 = 9.900"$$

The total tolerance on the whole stack would then be ±0.100", for a total of 0.200".

The problem with this analysis is that the odds of *all* parts being at the maximum or *all* parts being at the minimum are extremely low.

RSS TOLERANCE CALCULATIONS

Now, let's use the RSS (root sum-of-squares) approach to calculating tolerances. What should be the tolerance on each part if we assume that ±3 sigma (99.73 percent) of the products are within tolerance?

> ## Assume ±3 Sigma of Products Are Within Specification
> Unless you have specific data showing otherwise, assume that ±3 sigma (99.73 percent) of the products are within tolerance. This is a general rule of thumb, since few products are made totally within specification.
>
> TIP

Assume that in this example, we want ±3 sigma of the total stack height to be within the 10" ±0.100" tolerance specification, the same as assumed with the traditional approach to tolerances.

RSS: Calculating the Sigma *S* from Multiple Parts Stack-Up

$$S = \sqrt{(1.3s_1)^2 + (1.3s_2)^2 + (1.3s_3)^2 + \cdots}$$

S = the resultant assembly stack-up sigma

$s_1, s_2, s_3, \ldots$ = the sigma of each individual part being stacked

Each sigma s is multiplied times 1.3 to allow for a long-term sigma drift. If each of n stacked-up parts has the same sigma s, then:

$$S = \sqrt{n(1.3s)^2}$$

The 30 percent assumed long-term sigma drift ($0.3s$) included in the previous equation is based on actual data analyzed many years ago. It puts a degree of conservatism into the RSS tolerance method.

In our example, we first solve for the total stack sigma S. Assume that the ±0.100" tolerance, or 0.200", represents 6 sigma (±3 sigma).

$$S = 0.200"/6 = 0.0333"$$

This means that the sigma S on our total stack is 0.0333".

We then solve for the sigma s on each part, assuming that we will have the same tolerance on each of the 10 parts. We can solve for s from the following:

$$S = \sqrt{n(1.3s)^2} = \sqrt{10(1.3s)^2} = \sqrt{16.9s^2}$$

$$s = \frac{S}{\sqrt{16.9}}$$

$$s = \frac{0.0333"}{4.111}$$

$$s = 0.0081"$$

The tolerance on each part can therefore be 6 × 0.0081" = 0.049", rather than the 0.020" dictated by the worst-case tolerance method. This means that we could more than double the individual parts tolerance, with resultant potential parts savings!

Example, Part B

Another way to look at the problem is to assume that we keep the original 0.020" parts tolerance and see what kind of tolerance we can expect on the total stack assembly (again, assuming ±3 sigma parts within tolerance).

Solving for the parts sigma:

$$s = 0.020"/6 = 0.00333"$$

We now calculate the stack sigma:

$$S = \sqrt{n(1.3s)^2}$$
$$S = \sqrt{10(1.3 \times 0.00333")^2}$$
$$S = 0.0137"$$

Therefore, we can analyze the total stack with a nominal height = 10" and a sigma = 0.0137". The 6-sigma tolerance is $6 \times 0.0137" = 0.082"$, compared to the 0.200" assumed in the worst-case analysis. This means that anyone planning to use the stack can assume a more repetitive-sized stack with less than half the variation assumed with the worst-case analysis.

CASE STUDY: LOOSENING EXCESSIVELY TIGHT TOLERANCES

A machine had 160 multiple assemblies that opened and closed as the machine cycled. These individual assemblies had multiple cams, rollers, and other parts that contributed to a total variation in how much each assembly opened, which was critical. The tolerance of each of the parts that contributed to the assembly opening had been calculated using worst-case tolerance methods, since the stack-up of the tolerances was a concern. The resultant tolerances on some of these parts were 0.0003", causing the parts to be excessively expensive and almost impossible to manufacture.

When the parts tolerances were recalculated using RSS methods, the tolerances became 0.0008". Although these were still tight tolerances, the parts could now be made with standard machining methods at reasonable prices. Savings were more than $70,000 per year.

Problem 1

A company was building an insulator stack that consisted of alternating fiber disks and glass wafers, with a quantity of 10 each. The nominal thickness of the disk was 0.100", and the thickness of the glass wafer was 0.500". Therefore, the total nominal height of the insulator stack was:

$$\text{Nominal stack height} = 10 \times 0.100" + 10 \times 0.500" = 6.000"$$

The company had determined that it required the total height of the insulator stack to be between 5.960" and 6.040", so the stack's tolerance was 0.080". The company designer calculated tolerances using the worst-case design and divided the 0.080" tolerance equally among the 20 parts (0.004" each). Quotes were requested from various suppliers on the fiber disks and the glass wafers. The specification for the fiber disk was 0.098" to 0.102". The specification for the glass wafer was 0.498" to 0.502".

When the part quotes were received, the company was shocked to find that the glass wafers were going to cost five times what it had estimated. (The disk cost was about what was expected.) When it asked the suppliers why the glass wafer quote was so high, it was told that the 0.004" tolerance on the part required them to incorporate a grinding step after forming the glass. If they had a greater tolerance, they could eliminate the costly grinding step.

What would the tolerance be on the glass wafer and the fiber disk if the RSS method were used to calculate tolerances?

Answer: Assume that the tolerances on the wafer and the disk are identical and that the total assembly is to be manufactured with ±3 sigma of the insulator stacks within tolerance. Since the customer had specified that the total assembly must be between 5.960" and 6.040", the total tolerance is 0.080" and will be equal to 6 sigma.

1 sigma S on the assembly = 0.080"/6 = 0.0133"

$$S = \sqrt{n(1.3s)^2} = \sqrt{20(1.3s)^2} = \sqrt{33.8s^2}$$

$$s = \frac{S}{\sqrt{33.8}}$$

$$s = \frac{0.0133"}{\sqrt{33.8}}$$

$$s = 0.00229"$$

Therefore, the part tolerance could be 6 × 0.00229" = 0.014", rather than the 0.004" calculated by the worst-case method. This difference would enable eliminating the grinding step.

We have assumed in the previous examples that the sigma is based on using ±3 sigma of the tolerance. If you have historical data on actual parts dimensions, you can get even more accurate estimates of the proper tolerance. There are even software packages that allow you to put in raw data on the components. The software then fits a curve for each component, and you can do a Monte Carlo analysis, in which the software uses random numbers to generate data as if you were actually running thousands of parts. However, this is more sophisticated (and a lot more work) than is generally required.

Add Only Stack-Up Items on Axis
When using the RSS method of calculating tolerances, be careful to add only the parts' dimensions that are in the same axis as the stack-up dimension. If a part is at an angle relative to that axis, then you should include only the component that is in the axis direction.

TIP

Reviewing Tolerances on Real Parts and Processes
You are now ready to review tolerances on real parts and processes.
1. On any part that is difficult or expensive to make because of tight tolerances, review the tolerances based on the requirements of the application.
2. On any group of parts that are assembled in such a way that their accumulated stack-up dimensions affect the total assembly's dimension, review the tolerances using the RSS tolerance method.

TIP

WHAT WE HAVE LEARNED IN CHAPTER 17

1. Tolerances apply to the Improve and Control steps in the DMAIC process.
2. Tolerances are seldom calculated based on requirements, so there are potential savings to be realized by reviewing tolerances that cause issues.
3. When there is a problem with a part, the reaction is often to tighten the tolerance on that part, whether or not the tolerance is the issue.
4. If multiple parts are "stacked" in an assembly, the tolerances on those parts are likely to have been calculated using worst-case methods. Using the RSS method of calculating tolerances can open tolerances on those parts or show that the assembly has less variation than assumed.

RELATED READING AND SOFTWARE

Mikel J. Harry and Reigle Stewart, *Six Sigma Mechanical Design Tolerancing*, 2nd ed. (Schaumburg, IL: Motorola University Press, 1988).

MINITAB 13, Minitab Inc., State College, PA; www.minitab.com.

Crystal Ball 2000, Decisioneering Inc., Denver, CO; www.decisioneering .com.

CHAPTER 18

Simplified Linear Transfer Functions

W hat you will learn in this chapter is how to use simplified linear transfer functions to understand the effect of each component on the total variation of a part, an assembly, or a process. With this information, you will know each component's contribution to the total variation and know where to focus your attention. Simplified linear transfer functions are used in the Analyze and Improve steps of the DMAIC process. This chapter is primarily for those who are involved in design, manufacturing, or process work.

Just as with DOEs, there will be some people who think that this topic is too complex for this level of text. However, green belts *have* used this tool—and with great success.

The method used in simplified linear transfer functions is very similar to the RSS tolerance method that we already covered. In fact, this type of transfer function is called the *root sum-of-squares (RSS)* because it involves squaring and summing the contributing sigma figures, as we did for tolerance in the previous chapter. We want to account for all the variation in an assembly or a process by identifying the contributing variation of each component.

FORMULA

RSS Linear Transfer Functions

$$S_t = \sqrt{s_1^2 + s_2^2 + s_3^2 + s_4^2 \cdots}$$

S_t = the critical sigma of the total assembly or process

$s_1, s_2, s_3, s_4, \ldots$ = the sigma of the variables linearly affecting the critical sigma of an assembly or process.

Each variable's influence must be stated in common units that are consistent with the part, assembly, or process. For example, if we are studying the thickness variation of an injection-molded part and one of the contributing variables is the weight of the injected raw material, we need to state that variable's sigma in "thickness variation per sigma," rather than in "weight unit per sigma."

The best way to illustrate this is to tell about an actual project done in a manufacturing plant.

CASE STUDY: FINDING A GRINDING ISSUE USING THE TRANSFER FUNCTION

A manufacturing plant was manufacturing an item that involved pressing a forming material into a defined cavity. Exhibit 18-1 is a simplified diagram of this process.

Exhibit 18-1. Diagram of forming process

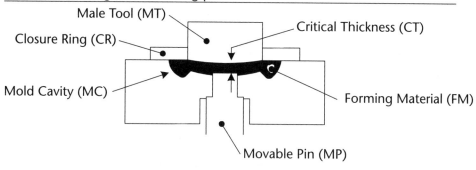

The critical thickness was varying too much, and the plant wanted to run tests to see what was causing this excessive variation.

First, the team defined the things that could affect this critical thickness:

S_{CT}: total sigma on the critical thickness

1. s_{FM}: sigma of the forming material
2. s_{MT}: sigma of the shape of the male tool
3. s_{MC}: sigma of the shape of the mold cavity
4. s_{MP}: sigma of the movable pin position
5. s_{AO}: sigma of all other unknown contributing variables

The formula for the total sigma S would therefore be:

$$S_{CT} = \sqrt{s_{FM}^2 + s_{MT}^2 + s_{MC}^2 + s_{MP}^2 + s_{AO}^2}$$

The team needed to run the tests in such a way that only one variable at a time was changed. The team members could then see the effect of each contributing sigma without the effect of the other variables. The only variable that they couldn't tightly control was the weight of the forming material, so it was decided to run that test first. Then, on tests for the other variables, they would analyze only product that had an average weight, eliminating weight as a variable on the remaining tests.

Test 1: Find the Sigma s_{FM} of the Forming Material Weight

Product was taken from one mold, where the pin was locked in position to eliminate any influence from the movable pin. Only one male tool was used. The weight was allowed to vary over its normal range. The critical thickness of the resultant product was measured. From these data, it was possible to calculate the effect of a weight change on the critical thickness, with no other variable changing.

There were historical data on weight variation, so the sigma in grams was already available. However, to get it into the common units needed, the team used the data previously given to convert the weight sigma to inches of critical thickness. This was a simple conversion, using the data on thickness to determine what the equivalent thickness change was at the gram-weight value of the historical-weight sigma. The team now had s_{FM} for the weight variation in critical thickness inches.

Test 2: Find the Sigma s_{MT} of the Male Tool

Product was taken from one mold, where the pin was locked in position to eliminate any influence from the movable pin. Multiple male tools were run. On products that had weights close to the historical average weight, the critical thickness was measured. In this way, the team was able to find s_{MT} in terms of critical thickness inches.

Test 3: Find the Sigma s_{MC} of the Mold Cavity

Product was taken from multiple molds, where the pins were locked in position to eliminate any influence from the movable pin. One male tool was used. On products that had weights close to the historical average weight, the critical thickness was measured. In this way, the team was able to find s_{MC} in terms of critical thickness inches.

Test 4: Find the Sigma s_{MP} of the Movable Pin

Product was taken from the same multiple molds used in the previous test, but the pins were no longer locked in position. One male tool was used. On products that had weights close to the historical average weight, the critical thickness was measured. The sigma was found for each individual mold. This gave a sigma for the movable pin in that mold. Using the RSS method, the sigma from all the molds was then calculated. This gave the effect of many movable pins. In this way, the team was able to find s_{MP} in terms of critical thickness inches.

The total S_{CT} of the critical thickness was already available from historical data. The s_{AO} of "all other unknown contributing variables" was the variable for which the plant then solved, since all the other variables were known and the S_{CT} equation had only one unknown. If the s_{AO} of the "all other unknown contributing variables" had not been small compared with the identified contributors, the team would have known that it had missed some important variable(s) and would have had to go back and review its understanding of the process.

Here is the resultant RSS equation:

$$S_{CT} = \sqrt{s_{FM}^2 + s_{MT}^2 + s_{MC}^2 + s_{MP}^2 + s_{AO}^2}$$

$$0.0071" = \sqrt{0.0015"^2 + 0.0046"^2 + 0.0013"^2 + 0.0049"^2 + 0.0011"^2}$$

When the team members analyzed the elements in the resultant RSS equation, they found that two variables were contributing most of the variation in the critical thickness. The highest, the movable pin at 0.0049", was no surprise, and the plant already had projects underway to correct this.

The sigma on the male tool, which at 0.0046" was almost as large as that on the movable pin, was a complete surprise. The male tool, whose surface was periodically reground in the plant, was varying far more than anyone had thought. It was discovered that the check procedures that were supposed to be used to verify that the grinding wheel shape for the male tool was correct were no longer being followed. This was quickly corrected, and almost a third of the problem of critical thickness variation was eliminated within a day, at almost no cost.

In solving this problem, the team took care to make sure that each sigma contribution related directly to the change in the critical thickness, with the units being consistent with the product effect being measured. As a result, comparing the sigma to see which variable was more critical was valid.

No attempt was made to detail the transfer function to the point that the formula for the "shape" of the male tool was included in the overall formula. In an ideal world, it would be nice to have the transfer function defined by the geometry and position of parts in space. It just is not normally required or practical!

Nonlinear Transfer Functions

TIP

There are some processes that are nonlinear and have such complex interactions among variables that they can't be represented with a simple linear transfer function. Chemical processes are often that way. The resultant transfer function is nonlinear and requires partial derivatives.

Tests to identify the components of nonlinear transfer functions are extensive, with a large number of test iterations required. This is beyond the scope of this book. Also, processes requiring nonlinear transfer functions are seldom defined completely.

WHAT WE HAVE LEARNED IN CHAPTER 18

1. Simplified linear transfer functions are used in the Analyze and Improve steps of the DMAIC process.
2. Use simplified linear transfer functions to understand the effect of each component on the total variation of an assembly or process.
3. The sum of the squares of the contributing variables' sigma must equal the square of the sigma of the total assembly or process. If the sum is too little, one or more variables are missing.
4. Each sigma contribution must have units that are consistent with the product effect being measured. When this is the case, it is valid to compare the sigma to see which variable is more critical.
5. Nonlinear transfer functions, which require partial derivatives, are beyond the scope of this book (and of most Six Sigma work).

RELATED SOFTWARE

Crystal Ball 2000, Decisioneering Inc., Denver, CO; www.decisioneering .com.

PART VII

Quality Department Data and Manufacturing Innovation

CHAPTER 19

Comparing Six Sigma Data with Quality Department Data

This chapter applies to someone who has to interface Six Sigma data with quality department data. Quality department data are often based on averages rather than on individual data points.

This chapter does not include a Six Sigma tool, but it *does* show how someone can use data collected by quality departments to compare with samples collected doing Six Sigma work. It is primarily for those involved in manufacturing or process work. What you will learn here is that quality department systems often use child distributions based on sample averages and the sigma of multiple sample averages, rather than using the individual data points.

To do statistical analysis, we use data in various ways. The previous problems in this book used the average and sigma of individual data points. This is called a *parent population*. Many quality departments deal primarily with sample averages and the sigma of multiple sample averages. This is called a *child distribution*. Quality departments often do not even retain the individual data measurements. One of the reasons this is done is to limit database memory requirements.

DEFINITION

Parent Population and Child Distribution

A *parent population* refers to the individual data and their related statistical descriptions, such as average and sigma. These are labeled $\overline{X}$ and S.

A *child distribution* refers to the sample averages and the sigma of multiple sample averages. These are labeled $\overline{x}$ and $s_{\overline{x}}$.

The average of the child distribution $\overline{x}$ will tend to match the average of the parent population $\overline{X}$.

The $s_{\overline{x}}$ calculated from multiple sample averages will be smaller than the population's sigma S. Note that the sigma $s_{\overline{x}}$ of multiple sample averages is *not* the same as the sigma s of individual data points.

The larger the size of the individual samples, the smaller the sigma $s_{\overline{x}}$ of the multiple means of the samples. You can't use the $s_{\overline{x}}$ of the multiple sample means as an indicator of the parent population sigma S without adjusting the $s_{\overline{x}}$ to account for the size of the individual samples. You can see in Exhibit 19-1 that the child distribution is "tighter" than the parent population.

Exhibit 19-1. Parent population distribution and child distribution

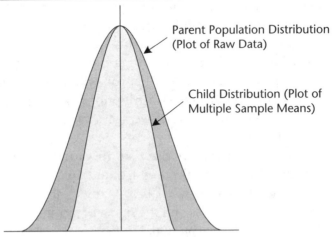

To estimate the parent population sigma S from the $s_{\overline{x}}$ of a child distribution, multiply the child $s_{\overline{x}}$ by the square root of n, the individual sample size.

FORMULA

Estimating the Parent Population Sigma S from the $s_{\bar{x}}$ of a Child Distribution

$$S = s_{\bar{x}}\sqrt{n}$$

S = parent population sigma (sigma based on the raw data)

$s_{\bar{x}}$ = child distribution sigma (sigma of the multiple sample averages)

n = individual sample size (quantity in each raw data sample)

As noted, the $s_{\bar{x}}$ resulting from multiple sample averages will be smaller than the parent population's sigma S. This is illustrated here by showing the shaft data for 15 samples, with each sample having 7 shafts ($n = 7$) from the parent population. The average of each sample of 7 was calculated. Then the sigma $s_{\bar{x}}$ (one value) of these 15 sample averages was calculated. To estimate the population sigma S from the sigma $s_{\bar{x}}$ of these 15 sample averages, multiply the sample average sigma $s_{\bar{x}}$ by $\sqrt{7}$.

This is shown in the following data. All dimensions are in inches. The samples were drawn randomly from the parent population of shafts, which has an average of 1.0000" and an $S = 0.0010"$.

Sample Diameters							
1	2	3	4	5	6	7	Average
1.0003	0.9994	0.9996	1.0012	1.0001	1.0004	0.9991	1.0000143
0.9982	1.0002	0.9989	0.9999	0.9996	0.9990	0.9980	0.9991143
0.9995	0.9999	1.0000	0.9976	0.9990	1.0005	1.0015	0.9997143
1.0022	0.9981	1.0003	0.9991	0.9994	1.0012	0.9987	0.9998571
1.0001	1.0008	0.9989	0.9994	0.9997	1.0018	0.9996	1.0000429
1.0014	0.9986	1.0009	1.0005	1.0010	0.9998	0.9977	0.9999857
1.0006	0.9996	0.9999	0.9995	0.9978	0.9990	1.0000	0.9994857
0.9995	1.0002	1.0015	0.9985	0.9998	1.0000	1.0007	1.0000286
1.0004	0.9992	1.0019	1.0009	0.9984	1.0010	0.9993	1.0001571
1.0016	0.9997	1.0007	0.9989	1.0003	1.0001	1.0011	1.0003429
1.0006	1.0020	1.0002	0.9994	0.9996	0.9986	1.0015	1.0002714
0.9991	1.0009	0.9992	1.0008	1.0006	1.0005	0.9996	1.0001000
1.0029	1.0001	0.9999	1.0012	1.0005	1.0007	1.0004	1.0008143
1.0023	0.9996	0.9997	1.0009	0.9990	0.9982	0.9996	0.9999000
1.0003	0.9991	0.9989	1.0008	0.9985	0.9998	1.0014	0.9998286
						Sample average $\bar{x}$ =	0.9999771
						Child sigma $s_{\bar{x}}$ =	0.0003850
Population standard deviation estimate = $0.000385 \times \sqrt{7} = 0.001018614$							

As you can see, the 0.999977" average $\bar{x}$ of these samples is close to the 1.0000" $\bar{X}$ of the population. The 0.000385" $s_{\bar{x}}$, however, is far smaller than the population S. As shown previously, we can approximate the population S (which is 0.0010") by multiplying the sigma of the sample averages $s_{\bar{x}}$ by the square root of n, or $\sqrt{7}$ (which is 2.646). This gives us an estimated $S = 0.001019"$, which is close to the population S of 0.0010".

Just for information, the actual average of the previous 105 individual data readings is 0.999977" and the sigma of the 105 raw readings is 0.001075". But remember, quality departments often do not keep the individual raw readings, so this comparison would not be possible on those systems.

Problem 1
Our now-familiar lathe is machining shafts, and, in this case, they are not averaging 1.0000" in diameter. We want to see whether an improved setup for the cutting tool will give us shafts closer to our 1.0000" target diameter.

Answer: We first plot individual readings from the initial process. We then compare this plot with the plot of individual data points after the change. Assume that we see that the general shape of the distribution curve has remained the same.

Before the process change, the quality department had been taking samples of 20 measurements. It calculated the average on each sample of 20 measurements and then discarded the individual data values. It kept a running child sigma $s_{\bar{x}}$ and child average $\bar{x}$ based on the last 100 of these calculated averages.

After the process change, we check a sample of 60 shafts. The following are the results before and after the process change.

Before Data	After Data
child $\bar{x} = 1.0004"$	sample $\bar{x} = 0.9999"$
child $s_{\bar{x}} = 0.0003801"$	sample $s = 0.00173"$
each child's $n = 20$	sample $n = 60$

What can we say about the average and sigma of the process "before" versus "after" the change?

The two sigmas can't be compared without some adjustment, because the "before" is a child sigma. However, we can estimate the population S from the $s_{\bar{x}}$.

$$S = s_{\bar{x}}\sqrt{n}$$

S = estimate of the parent population sigma of the "before" process
$s_{\bar{x}} = 0.0003801"$

$n = 20$
$S = 0.00170"$

Use the formula from Chapter 12 to see whether the sigma are significantly different.

Chi-squared test of a sample sigma s versus a population sigma S:

$$\text{Chi}_t^2 = \frac{(n-1)s^2}{S^2}$$

$n = 60$
$s = 0.00173"$
$S = 0.00170"$ (estimated from the child)

$$\text{Chi}_t^2 = \frac{(60-1)0.00173^2}{0.00170^2}$$
$$\text{Chi}_t^2 = 61.1007$$

We compare the calculated Chi_t^2 results with the values on the simplified chi-squared distribution table (Exhibit 12-1). If the Chi_t^2 test value we calculated is less than the table low value or greater than the table high value, we are 95 percent confident that the sample sigma s is different from the sigma S of the population. Since 61.1007 is between the table values of 39.662 and 82.117, we can't say with 95 percent confidence that the sigmas are different.

We now test the averages. Since we are interested in which average is closer to the 1.0000" nominal, we will restate the averages versus the nominal as deltas: $|\overline{X} - 1.0000"|$ and $|\overline{x} - 1.0000"|$

Before Delta
$\overline{X} = 0.0004"$

After Delta
$\overline{x} = 0.0001"$
$s = 0.00173"$
$n = 60$

t test of a population average $\overline{X}$ versus a sample average $\overline{x}$:

$$t_t = \frac{|\overline{x} - \overline{X}|}{\dfrac{s}{\sqrt{n}}}$$

$\overline{X} = 0.0004"$
$\overline{x} = 0.0001"$
$s = 0.00173"$
$n = 60$

$$t_t = \frac{|0.0001 - 0.0004|}{\frac{0.00173}{\sqrt{60}}}$$

$$t_t = 1.343$$

We then compare this calculated t-test (t_t) value with the value in the simplified t table (Exhibit 12-2). If our calculated t-test (t_t) value is greater than the value in the table, then we are 95 percent confident that the sample is significantly different from the population.

Since 1.343 is not greater than the table value of 2.001, we can't say with 95 percent confidence that the changed tooling setup is different from the original process.

The Child Distribution Tends to Be Normal

When you plot the averages of multiple samples taken from a parent distribution of any shape, the distribution of the child population tends to be normal. This is especially true when individual sample sizes are 30 or more.

This is known as the *central limit theory*.

TIP

When we looked at nonnormal distributions, we gave the examples shown in Exhibit 19-2.

Exhibit 19-2. Nonnormal distributions

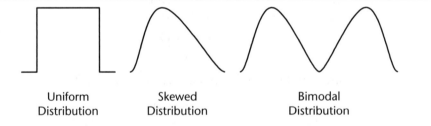

| Uniform Distribution | Skewed Distribution | Bimodal Distribution |

If you took multiple samples of 30 from any of these populations and plotted the means of these multiple samples, the resultant child distribution (Exhibit 19-3) would be normal. In fact, from the shape of the child distribution, you could not guess the shape of the original parent population.

Exhibit 19-3. Plot of the means, with samples of 30, from the nonnormal distributions in Exhibit 19-2

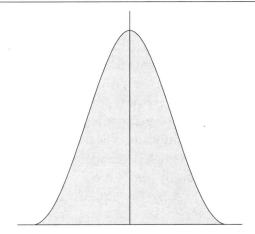

Be Careful When Using a Child Distribution

1. Since you can't tell the shape of the original parent population by looking at the child distribution (the plot of multiple sample means tends to be normal no matter what the parent distribution is), the parent distribution could change and you would not know it. You must periodically plot the raw data from the parent population to make sure the process has not changed.
2. Don't ever compare a child distribution sigma with a parent population sigma without first adjusting the child sigma by multiplying it by the square root of the individual child sample size.
3. Don't be misled into thinking that a process is in control because the sigma of the child population is small. The *real* sigma of the parent population is always larger.

Be Careful When Comparing Short-Term Data with Long-Term Data

When you are running tests, the data you collect are generally short-term (several days or less). However, quality department data can contain both short-term and long-term data. Long-term data include some process drift, which will increase the sigma value. The rule of thumb is to assume a 1.5-sigma increase because of long-term process drift.

However, for your data comparisons, you are safer if you use recent data with a time period equivalent to your test data, rather than assume a long-term drift value. This is because the actual long-term drift value can vary widely.

WHAT WE HAVE LEARNED IN CHAPTER 19

1. This chapter does not include a Six Sigma tool, but it *does* cover how someone can use the type of data collected by many quality departments to compare with samples taken related to process trials or samples that are collected doing Six Sigma work.
2. Quality department systems often use child distributions, which are based on multiple sample averages, rather than on the raw data themselves. The raw data are often discarded after the average is calculated.
3. The sigma of the child distribution is always smaller than the sigma of the parent population, which is calculated using the raw data.
4. An estimate of the parent population sigma can be made by multiplying the child sigma by the square root of the individual child sample size.
5. The child distribution tends to be normal no matter what the shape of the parent distribution. This is especially true with individual sample sizes of 30 or over.
6. You have to be careful when using a child distribution. It can hide a change in the shape in the parent population, thereby hiding a process change. Also, the child sigma can't be compared with a parent population sigma without first adjusting the child sigma with reference to its sample size.

RELATED READING

Mark J. Kiemele, Stephen R. Schmidt, and Ronald J. Berdine, *Basic Statistics: Tools for Continuous Improvement*, 4th ed. (Colorado Springs, CO: Air Academy Press, 1997).

CHAPTER 20

The Next Step
for Six Sigma:
Manufacturing Innovation

Manufacturing innovation is currently not part of Six Sigma. But since innovators often use Six Sigma or similar analysis techniques in their work, and since manufacturing innovation is sorely needed in our current economy, it is important for practitioners of Six Sigma to understand what role this methodology can play and how it can assist them in the innovation process.

TRAITS OF MANUFACTURING INNOVATORS

It is very difficult to teach innovation. In fact, few universities even try. But it is possible to look at some of the traits of innovators and to study case studies of actual manufacturing innovations. From these studies, some lessons can be learned. Here are some common traits of manufacturing innovators that the author gleaned by studying 37 different manufacturing innovations in 6 different manufacturing plants.

Besides general engineering skills, successful manufacturing innovators have:

- A high sense of confidence bordering on ego
- Little fear of authority—everything is questioned

- The ability to work around or through barriers
- Persistence to the point that serendipity starts to be evident
- An insistence on excellence
- The ability to work with hourly people—to get their trust and build on their intimate observations of the process
- The ability to "see" a problem and its related opportunity with more focus and clarity than most people
- The ability to recognize when available solutions can apply to unique areas

Managers can look for people with these characteristics if they want to build an innovative production environment. Equally, managers can make sure that barriers to innovation, which are discussed in this chapter, are kept to a minimum. Engineers can see which of these traits they already have and build on areas where they are weak. Or by looking at these traits, they can see if they even have the aptitude or the desire to be a manufacturing innovator.

MANUFACTURING INNOVATION IS VERY DIFFERENT FROM PRODUCT INNOVATION

If you read about true product innovators, such as Steve Jobs at Apple Computer or Bill Gates at Microsoft, you will see behaviors, attitudes, and actions that could never work in an existing manufacturing facility. In both cases, Jobs and Gates started out with rather clean slates, with few limitations on how they approached their product design issues. They didn't have existing facilities, equipment, processes, philosophies, products, and workforces that they had to work around. There weren't any experts, management, and unions that they had to convince.

Not only did Steve Jobs and Bill Gates work long hours in pursuing their ambitions, but they surrounded themselves with like-minded individuals. They set up goals that some would consider unrealistic, yet they could insist that others around them pursue those goals. They didn't have to prove their ways to other people.

Both Jobs and Gates, at times, treated their employees with disdain. As discussed in Carmine Gallo's book *The Innovation Secrets of Steve Jobs*, Jobs would regularly call employees "bozos" if he felt that they were trying to push through an idea that was not in the best interest of the customer. Of course, Jobs was the one who decided whether an idea was in the best interest of the customer. In the book *Hard Drive: Bill Gates and the Making of the*

Microsoft Empire, by James Wallace and Jim Erickson, the authors note how Gates's abrasive personality led to his being labeled "The Silicon Bully" by *Business Month* magazine. And some of Microsoft's aggressive practices, even if not technically illegal, were often called unethical by competitors.

Anyone who tried to emulate Jobs's and Gates's personalities in an existing manufacturing facility would have issues to deal with. Employees with that kind of aggressive nature would not have been hired in the first place. Thus, many employees who had to deal with people of this type would either quit or have to be fired. Most ongoing manufacturing facilities could not afford such a risk; nor would their existing customers accept the resultant supply interruptions.

True manufacturing innovation requires incorporating much of the current manufacturing environment and personality into the new order. Sure, there can be some adjustment of employees, equipment, and products. But these adjustments must normally be done on the fly, with minimal costs and minimal disruption to any ongoing processes. And they must be done in an environment that is not geared for innovation and may indeed be antagonistic to change.

Product innovators working at start-ups generally know that there is a small chance that they could strike it rich if their product really takes off. Not so for manufacturing innovators, especially those in large corporations like GE or General Motors. If they have a manager who takes kindly to innovation, they may get a good review, a small stipend, or even a few stock options. But generally, manufacturing innovators don't get huge financial rewards. And few of them get to the vice president level in large corporations. Innovators' ways are often at odds with a large corporation's standards, and innovators often ruffle their managements' feathers.

Common Traits of Product and Manufacturing Innovators

Most innovators are willing to work long hours to make their ideas work. And these hours are often in excess of the hours they are already working for an ongoing job, since at the start of an innovative idea, much groundwork and modeling is done informally and not as an assigned project.

Another common trait is perseverance. In 1995, in an interview with the Smithsonian Oral History Project, Steve Jobs noted that perseverance is critical. If someone backs down the minute anyone questions a new way of doing something, or if someone gives up at the first sign of failure or backs off when the solution doesn't come together as smoothly as he expected, he will not be a successful innovator.

Innovators of all types generally have large egos. They have to believe in themselves, and they have the audacity to think that they can do something better than others can. There are innovators in Silicon Valley who have worked on many totally unrelated products and yet have managed to develop breakthrough products that were eventually sold to large corporations. There are rather inexperienced engineers who have looked at practices that had been in place for many years and proved them wrong. And there are bright hourly employees with no education beyond high school who innovated in ways that are difficult to understand. In all cases, these innovative people had high self-confidences and large egos.

Innovators also have an ability to truly see! All innovators can look at an area of promise or an area with problems with a focus that far exceeds what most other people can muster. And if an innovator is interrupted while she is doing this, she may not be very civil. Any innovator, especially in a manufacturing environment, knows that you need to incorporate the best computer (your brain) and the best sensors (your eyes, ears, and nose) if you are going to maximize the ability to truly see and understand.

Innovators settle for nothing less than excellence. Again, this is a quality attributed to Steve Jobs. For an innovator, this means excellence in clarity of thought. It means excellence in understanding an issue or an opportunity. It means excellence in building a model, in testing, in implementation, and in follow-through. I saw one innovator put in a truly remarkable piece of equipment on a Friday that literally eliminated a defect. He and several other engineers stopped by the manufacturing plant on Saturday morning just to check on how the equipment was running before they headed home. They saw a few defects. Although the level of defects was very low relative to what it had been before the engineer installed his project, he did not leave that plant on Saturday until he truly understood what had changed. He was eventually able to get the project back to 100 percent efficiency. While he was doing this, the engineers waiting for him had a lot of frustration. He ignored them because he *had* to understand and solve the issue that was hurting his project. Anything else would be accepting less than excellence!

Manufacturing Innovation Is *Not* Just Continuous Improvement

A philosophy of continuous improvement, a practice that is followed vigorously in Japan and in some companies in the United States, involves continually tweaking a process to achieve better quality or more productivity.

It is an ongoing implementation of many small ideas, often coming from a myriad of people. It is a bottom-up approach to improving a process or product. Over time, the cumulative improvement resulting from these changes can be significant. But as great as continuous improvement is, it seldom results in a substantial change in the original DNA of the manufacturing process. Generally, that degree of change requires a quantum step that comes from innovation.

Manufacturing Innovators Often Utilize Six Sigma–Type Tools

Six Sigma is a way of identifying problem areas and keeping score—knowing the severity of a problem or measuring whether a targeted change is having the desired effect. Although the Six Sigma methodology seldom leads directly to innovation, Six Sigma includes many tools that manufacturing innovators use.

The Six Sigma tools that are likely to be used include the QFD and its complementary Six Sigma tool, the FMEA. Fishbone diagrams, process flow diagrams, and visual correlation tests can assist in identifying a problem and a possible innovative solution. Gauge verification is required in order to have confidence that any measurement error is not excessive, so that the data can be believed. And plotting and statistical tests of data are required in order to truly understand many problems before potential manufacturing innovations can even be considered.

SO, WHAT EXACTLY *IS* MANUFACTURING INNOVATION?

Manufacturing innovation is finding a breakthrough change in a manufacturing process step that takes the process to a higher level or opens up new manufacturing horizons.

The goal of manufacturing innovation is not just getting a small improvement—it is an attempt to radically change a whole manufacturing process step, or even eliminate that step. After an innovation solves an ongoing problem, a production plant can often speed up or go after new business.

Unlike product innovators, manufacturing innovators often rely on those working directly in the processes, who are often hourly employees, to trigger insights. These contributing people must be treated with respect. As you will see in many of the case studies in this book, employees who are working directly in a process often sense that some major change is needed. They may not know the exact change required or the physics behind a potential change, and they may not have any data supporting their ideas. And yet, their intimate familiarity with the feel of the process gives them some degree of confidence that there is a breakthrough just waiting to be found and incorporated. They just need some help in identifying and incorporating the details of that breakthrough. That is where innovators step in.

This interaction between the innovator and someone who is directly involved in the manufacturing process is one of the reasons that manufacturing innovators are not plentiful. Innovators of all kinds are often introverts who like to work alone and don't always play well with those who are less technically inclined. And yet, in a manufacturing environment, a connection between the process people and the innovator is often required. Interestingly, once an innovator is able to cross that people interface threshold, these relationships often turn into extremely pleasant work teams, with respect going both ways. In the case studies in this chapter, you will see many instances of this.

Standing in the way of manufacturing innovators are experts, historical beliefs, unions, managers, quarterly profit targets, facilities, customers, "it's been tried before," and an overall fear-based aversion to major change. In this chapter's case studies, you will see examples of many of these barriers—and you will see how they were worked around. When reading the case studies, you should note the generic nature of the problem addressed, not the detailed specifics of the solution. It is unlikely that you will encounter the exact same problem discussed in the case study in another manufacturing plant. But recognizing the path to the solution, and noting how the barriers were surmounted, will be valuable as you attack a plant's specific problems. History doesn't repeat. But it certainly does rhyme!

Let's discuss the barriers to manufacturing innovation further. Experts are often a huge barrier. There are three categories of experts. The first is the general expert who has been in a manufacturing plant for many years and has knowledge of and responsibility for a wide range of plant activities. Another kind of expert is the person who feels that she has a very deep and complete understanding of a very specific technical area, with many years of experience backing up this expertise. A third kind of expert is someone who takes over though sheer willpower, dominating and micromanaging everything. The ideal way to deal with experts is to make them feel that they are part of any solution. If this is not possible, then an overwhelming amount of historical or test data must be collected to prove the expert wrong.

Historical beliefs or practices are also barriers because beliefs can be entrenched in many people's minds and in many procedures. No matter how much data or logic a manufacturing innovator accumulates, there will be those who will want to cling to past beliefs. The only way to deal with this kind of barrier is to make any change as quantitative as possible and have it incorporated into an equipment change or procedure modification that does not require the active acceptance of everyone. A qualitative change that requires the cooperation of all those using the new concepts will require constant monitoring to make sure that people don't slowly drift back to prior practices.

Unions can be either a barrier or a source of support. If union leaders can be convinced that a proposed change will make the employees' jobs easier or more secure, or will improve a product, then the union can be an active support factor. If a change, no matter how logical or how good for the company, involves people reductions, then the union is far more likely to fight the change. However, if the people reductions can be made concurrent with retirements or if the incentives to retire early are attractive enough, then the union may indeed support the change.

If a union is very militant and is constantly fighting management, then no matter what the innovator does, there could be union issues. Even if the innovation includes adding people, there could be issues of seniority, work practices, or the pay rate for the additional jobs. But thinking of these issues before the change is implemented can greatly reduce conflict.

Management can be as much of a hindrance to change as unions can. In many companies, quarterly targets are sacrosanct, and anything like experimental trials that might put those targets at risk will not be supported. Managers know that any change, no matter how well thought out, has risks. And some managers are risk-averse by nature. If a manager who is risk-averse is the innovator's direct supervisor, the innovator may very well be blocked from making any large changes. I knew of one proven innovator who got a new boss who was risk-averse. This new boss not only gave the innovator clear instructions to cease and desist, but also gave him a negative review for his past successful innovations. The manager told the innovator that he had been putting the company at risk with his prior projects, no matter how successful they were. The innovator's only option was to move to a different part of the company where he could work for a different manager. In the worst-case scenario, he would have had to leave the company.

Existing facilities are a concern for manufacturing innovators. No matter what the breakthrough idea, manufacturing plants are unlikely to throw out their existing equipment and start over. When considering the means of innovation in a production facility, the successful innovator will try to piggyback on or modify current equipment at minimal cost and risk.

Customers can also be a barrier to innovation. Any change in a process can cause unintentional and imperceptible changes in a customer's product. Any change in that component could cause issues that are not recognized for many months, putting the customer at huge risk. The use of the Six Sigma FMEA can minimize risk, but it cannot eliminate it completely.

This kind of delayed problem happened in a company making household lamps. An engineer proposed changing to a less expensive material for the filament of a standard household lamp. Since lamps burn for many hours, normal life tests are difficult. So life tests on the new filament material were done with high voltages and then extrapolated to the total life of the lamp under regular use. These tests showed that the new filament material would not affect the life of the lamp. Therefore, the change was incorporated. However, the conversion tables for high-voltage tests were all done using the prior filament material, and they apparently were not valid with the new material. It took the company several years to realize that it had shortened the life of its lamps by about 10 percent. It was forced to go back to the former filament material. Interestingly, in this case, the company actually made money on this error, since it sold more lamps during those several years. It was the consumers that paid the extra millions of dollars because the lamps didn't last as long!

If a change is made to a process making a product that is sold directly to a consumer, then feedback on the resulting issues can be very difficult to collect. It could be that by the time a problem is identified, the customer is already moving to a competitor. The company that was making household lamps was very lucky that it found and corrected the short-life lamp problem before either the customer or the government recognized the issue. This episode also underscores why some managers are so resistant to change.

The "it's already been tried" reason for not supporting innovation is often a barrier. Even though the earlier trials on an innovation may not have been exactly the same as the change that is being proposed, the person doing those earlier trials was probably just as confident of their successful outcome. This sours people on the likelihood of any additional attempts being successful.

Here are some case studies of manufacturing innovations that took a manufacturing plant to higher levels in both quality and production. These first three case studies all took place in the same manufacturing plant, but involved different parts of the manufacturing process. One innovation involved the uniformity of the raw material being supplied to the process. Another involved a process change that eliminated a defect, and the last innovation involved automatic inspection of a defect that was critical yet very difficult to inspect for.

CASE STUDY: DIFFICULTY IN MAKING STAINABLE LENSES

Glass plants that make high-volume glassware for uses like headlamps or floodlamps normally don't achieve the pristine glass quality of a small glassmaker. It would require far more costly procedures and raw materials, and much slower speeds to reduce the pull on the glass-melting furnace, to get that kind of uniform glass. And yet, GE made several types of clear lenses that had to be stained (for uses like school bus lights). These lenses required very high-quality glass without cords. (Cords are strands of slightly dissimilar glass within the lens. These glass cords are often invisible, but they show up dramatically when the glass is stained.) Customers would not accept stained glass with cords.

There was a senior engineer who had spent years trying to get glass that was stainable. He would take the production line that was going to run the glass out of production several days early so that he could precondition the glass. He would overheat the glass coming out of that line so that it would melt away any old glass. He would also treat the forehearth (the ceramic tunnel leading from the main furnace to the production line) chemically for the same purpose. Then he would slowly bring the glass back to temperature. He would then run the stained types immediately. The problem was that this procedure worked only about 50 percent of the time, so the plant would regularly run out of stainable glass. This would really irritate the customer, who couldn't understand why the glass plant did not run a zillion lenses when the glass was stainable and put them into inventory. This was not a workable solution because even when the glass was acceptable for staining, it would deteriorate within a day or two, precluding any extended run for inventory build.

There was a very creative press operator on the production line who suggested to the engineer that he modify the ceramic ring through which the glass flowed so that it had a lip that stuck up into the glass stream. The operator felt that this would solve the problem. He had been suggesting this for some time, but the engineer concluded that the most this would do was buy a few hours until the nonflowing glass built up to the level of the lip. The engineer thought that this would just work as a temporary dam.

The engineer wasn't having much luck solving the glass issue, however, so out of frustration, he decided to give this lipped ceramic ring, which he called a spout, a try. After getting the ceramic spout made, he put it in, and he instantly had good stainable glass. This didn't surprise him because he had expected that this would happen for a few hours, until the unmoving glass around the spout built up to the lip level. But 8 hours later, the glass was still good; 24 hours later, the glass was still good; 48, 72, 96 hours later, the glass was still good, and there was no sign of glass quality deterioration.

Since no one had expected a continuous improvement in glass quality, something was happening here that no one had expected. After much discussion, the engineer and the operator concluded that the lip on the spout affected not only where the glass was coming from at the point the glass entered the spout, but also the total location of the glass stream in the forehearth before the glass even got to the spout. The spout apparently put the glass stream in an area that was removed from the ceramic forehearth walls that contained the glass: a sweet zone. The plant was so happy with the results that it eventually used a spout on all types of glass because a more uniform glass was likely to help in the processing of all glass types.

This senior engineer did not have a history of innovation. But in this case, the engineer and the operator became innovators by dramatically improving the homogeneity of the raw material coming to the pressing process.

To show how remarkable this breakthrough was, several years after the incorporation of spouts, engineers from Corning Glass visited the GE plant as part of a technology exchange. Corning Glass had previously been a competitor of GE, but Corning was getting out of the lamp glass business to concentrate on other more profitable parts of its business, like fiber optics glass.

Corning Glass is probably the most advanced glassmaker in the United States and certainly up with the top few in the world. But when the GE glass plant had been having trouble getting stainable glass a few years earlier, Corning had also been having problems getting stainable glass. GE knew this because Corning would call GE periodically to see if GE could supply stainable glass, just as GE would call Corning for supply when GE was having problems. However, Corning solved its stainable glass problem at just about the same time GE did. GE knew this because it was no longer getting supply calls from Corning.

Corning Glass put far more engineering and technological support into its glassmaking than GE did. In fact, on all its shifts, Corning had an engineer whose only function was to worry about glass quality. GE had no such person on any shift. Thus, the GE engineers were curious about what Corning would think of the GE plant when they saw it.

The Corning engineers were amazed. The GE engineers thought that Corning would be derisive of the plant for its relatively crude ways. Instead, the Corning engineers were very complimentary that GE had been able to generally match Corning's glass quality with far simpler solutions, the spout being one of them. To solve its stainable glass issue, Corning had coated all the ceramics in contact with the glass with platinum, one of the most expensive metals in the world. It had also installed a platinum stirring device right over the glass exit to keep the glass homogeneous. The engineers couldn't believe that the simple addition of a lip on the glass exit ceramic ring did the same thing!

The hourly person who suggested the spout got the largest suggestion award ever given by that glass plant. He deserved it!

The barrier to this project was GE's having no understanding of how the glass really flowed from the furnace. The solution was the engineer finally trying what the press operator had been suggesting and experiencing the improved glass. Only then could GE get some understanding of what was apparently happening with the glass flow.

The spout innovation by this hourly worker was an example of an innovation idea coming as a result of an experienced and excellent hourly operator's feel for the process rather than through any detailed engineering study. And the innovation became possible only because of the senior engineer's willingness to let a far less educated person, and someone with absolutely no authority or expertise in glass melting, influence him enough that he designed and had built a spout to try!

There was another GE glass-pressing plant that refused to even try the spout. Since the other plant did not make clear glass that had to be stained, it had no obvious need for it, and the engineers at the other plant refused to accept the idea that getting more uniform glass was very likely to help their customers on *all* glass issues. The "not invented here" issue probably influenced this, since the two glass plants competed with each other in many ways.

The next case study involves the glass-pressing operation itself. This innovation was done in the same plant that made the breakthrough on the spout. This is typical: if a plant has management that encourages innovation, success creates an atmosphere that encourages even *more* innovation.

CASE STUDY: LENS WARP DUE TO RIBS

Some pressed glass lenses, rather than being of uniform thickness, have directional ribs going across their inner surfaces. This is especially true for lenses that have optical purposes other than uniform lighting. These lenses tend to have a high degree of warp.

Traditionally, in many pressed glass plants, operators tried to reduce the warp of their ribbed lenses by putting flat plates on the press that would push down on the rims of the lenses several stations after pressing. The idea was to push the rim of the glass back down against the flat rims of the molds to take out any warp. And there were always arguments about which position on the press was best to do this. Some operators would try to have the plates hit the glass at an angle; others would use plates that were not flat so that they could get more directed pressure in warp areas. And of course, all of this had to be done gently so that no defects would be put into the glass rim. But one questioning engineer never saw any of this reduce warp in a consistent or measurable way!

The questioning and innovative engineer looked for a more effective solution. After lenses are pressed, overhead cooling air ducts blow air down on the glass so that the lenses will cool enough to be taken out of the mold some stations later. The engineer believed that since the ribs on the inner surface of the lens were most exposed to this overhead cooling, they cooled first. When the glass cooled, it first would get rigid, then grow smaller as it cooled further. Since the ribs became rigid and got smaller long before the hot surrounding lens glass did, as the hotter remaining glass eventually became rigid and got smaller, it curled around the already rigid ribs (because the ribs had already cooled and therefore could not get smaller). This theory was supported by the warp direction correlating with the rib direction!

To compensate for this warp, the engineer took a high-volume lens type that had directional warp problems and measured how great, on the average, the directional warp was. It was 0.008" at 90° away from the rib direction.

A machined ring that is around the plunger forms the rim of the glass during pressing. Therefore, the engineer had a ring specially machined such that it was gradually 0.008" deeper in the two areas of the glass that were *not* experiencing warp. These were the areas at the ends of the directional ribs. Using this modified ring, the formed lens had an extra 0.008" inch of glass in those two areas. As the lens warped, it warped enough so that the average lens surface was flat once it had cooled. Any warp that was left was nondirectional and was well within the customers' specifications.

This empirical correction for warp was such a unique solution that it received a patent. Also, in a technical exchange agreement with another glass company that made huge lenses for glass TV screens, this concept was part of the technical exchange because the other glass company had so much trouble with nonflatness that every TV lens it pressed had to be ground. This nonflatness was not random; it was directional. This grinding process was a very expensive and time-consuming operation. The company making these TV lenses hoped that with the empirical rim correction idea, the amount of required grinding would be greatly reduced. And it was!

The barrier in this case was that everyone kept trying to reduce warp on the press by using plates to push the rim down after pressing, even though this solution wasn't working. The other barrier was that people had a hard time accepting that warp was an inherent part of the ribbed lens design rather than being a process problem. Rather than fighting this natural warp tendency, the innovative engineer expected it and compensated for it empirically!

Einstein's definition of insanity is doing the same thing over and over and expecting different results. The innovator saw this going on with the plates on the press being used to reduce warp. The engineer used Six Sigma visual correlation tests to show how the warp direction was correlated with the rib direction. He then empirically corrected for it.

The next case study is about a project that affected the finished product inspection. Again, this was in the same plant that did the spout and warp correction. The plant's people were very innovative, and they got the required support from management.

CASE STUDY: INSPECTING FOR RIM CHECKS

Ever since GE had started to make its own glass, rim checks had been an unsolvable problem, both in terms of their cause and in terms of the inspection methods used to try to inspect for them. But first, let's give a definition of rim checks. Rim checks are extremely fine cracks in the rim of the glass. Many of these checks are narrower than a human hair, and the only way you can find them as you are inspecting the glass is to keep changing the angle of the glass until a light source flashes off the plane of the check going into the rim of the glass. Light bouncing off this check plane will cause a momentary flash that will catch the inspector's eye. However, the angle of the glass has to be just right for this to happen. Obviously, at press speeds of up to two pieces of glass per second, manually inspecting for this defect was an impossible task!

A small rim check would sometimes grow into a crack when it was heated in the lamp-making process, and then the glass would break. When the glass broke, it would sometimes cause jamming of the lamp-making equipment and much downtime. Therefore, this was a critical defect.

Over the years, many things were tried to keep these checks from getting to the lamp plant. One of the trials involved quickly heating, then cooling the lens in an attempt to get the check to break. However, trials showed that this process had to be repeated many times to eliminate this defect, and there was no room in the glass plant to carry out such a time-consuming process.

Engineers also tried to use sound. The thought was that high-frequency vibrations would break any glass that had a rim check or cause the glass to vibrate in a mode that was identifiable. That was not successful.

Engineers tried physically tapping the glass in various positions during the process to see if that would cause the rim check to break. Unless it was smacked hard enough that many good parts broke, this did not work.

The plant could slow down the pressing operation; that would reduce the incidence of checks. But even that did not completely eliminate rim checks, and a slowdown was very costly. As for automatically inspecting for rim checks, even though the engineers had put automatic inspection devices in the production lines to look for all the other major defects, no one knew how to automatically inspect for rim checks.

One day the manager of the customer lamp plant that received this glass came and talked to the glass plant's engineering team. He wanted to talk about automatic inspection for rim checks. He acknowledged that many people had looked at this problem and that no one had succeeded in solving it. But since this team had successfully managed to automatically inspect for all the other major defects, he requested that the team give rim checks another shot as a personal favor to him. He acknowledged that the rim check problem had no obvious solution; but he had been impressed with the team's innovative work on other difficult auto-inspection challenges, and he truly respected this team's abilities to solve tough problems. The team members promised him that they would try!

To keep this issue in their minds, the lamp plant manager handed each team member several lenses and reflectors. Each of these lenses and reflectors had a small rim check. He asked the engineers to keep these lenses and reflectors on their desks to remind them to periodically think about this problem of rim checks.

It so happened that about a week later, one of the more innovative engineers was waiting for another engineer when he noticed the problematic pieces of glass on the desk. There also happened to be a small laser pointer on the desk. So the engineer started to play with the laser and the glass.

The engineer knew that light can pipe around a curved glass surface, so he was just playing with the laser on the rim of the lens, shining it parallel to and in line with the rim. To his surprise, the light not only piped around, but also caused the whole plane of the rim check to "glow" if you looked at the lens from above. And it appeared that the point where the laser beam entered the rim could be quite a distance away from the check circumferentially, and the check still glowed. With the laser lighting the rim, when you looked from the top of the lens, it wasn't just a thin line that you saw. There was a substantial width to the glowing rim check area. The engineer did some playing to see how far away circumferentially the laser beam could be from the check, and it looked as if it could easily be 90 degrees away and still cause the check to glow substantially. There were several more lenses and reflectors on the engineer's desk, so he tried the same thing on those other checks. The checks glowed easily!

The other engineer was back by then, so together they went to look for more laser pointers. They found the four they needed. They cleared off an area on the desk and taped down several pieces of cardboard to use as a track to slide the glass through. They then taped the four laser pointers at 45° angles to this track. These made a square light beam pattern the size of the glass, with the laser beams hitting the glass at 90° intervals.

When they slid a lens or reflector through this track, there was one point in its travel when the rim was hit with the light from all four laser pointers at the same time. The engineers then played with each piece of glass to verify that no matter what the rotational position of the glass, the check would glow when it was looked at from above. And areas where there were no checks did *not* glow, which was equally important. They had just found an innovative way to inspect for rim checks!

They knew that if they did this same thing on a conveyor belt, the lenses could go by four laser beams continuously at high speed, and a digital camera mounted above could send the images to a computer. Any rim check would glow and could be identified in the image by a computer program. The output of this computer program could then cause the bad lens to be rejected from the line with a pulse of air.

The engineers wanted to first build a simple model and test hundreds of lenses and reflectors to see if they all acted the same way. Now, at this point in a project, you can get a lot of enthusiastic support. One engineer on the team volunteered her children to build the model. She had two boys who were in high school and who planned to go into engineering. She thought they would enjoy being part of a project at this stage and would learn from it. They had woodworking equipment at home, so building a model would not be hard for them to do.

In a few days, she came back with a wooden model. Her children had the lasers located and had mounted an overhead structure with a seeing hole over the center of the four lasers. This was so that someone could look down and see exactly what the digital camera would eventually see.

Then the engineers started pushing hundreds of lenses and reflectors through this model to see if the rim checks, and only the rim checks, glowed from the lasers. This was not successful! Some of the glass also glowed at a point called the inner ledge. However, the diameter of the inner ledge circle was substantially less than the outside diameter of the lens, so the engineers realized that they would be able to build an electronic mask within the camera/computer logic so that it would ignore any light flash coming from somewhere other than the outside rim of the glass. Other than the inner ledge flashes, the model showed that the concept looked promising.

The engineers were now ready to ask for some funding to build a more elaborate model that replicated what they thought the production unit would look like. The advanced model would include a moving conveyor belt to bring the glass in front of the lasers at a uniform speed, and a digital camera with logic that would allow them to incorporate the electronic mask so that the camera would ignore the area where they were getting some spurious glow.

Using the wooden model as a sales pitch, the engineers were then able to get funding (they also found some money to pay the students who had built the wooden model). They had an engineering firm build the advanced model. The inspection by this new model was 100 percent effective on rim checks, with no false positives—no good lenses or reflectors were called bad. For their tests, they had used Six Sigma statistical methods to determine the minimum number of lenses with and without rim checks that they needed to run through the equipment to be 99.9 percent sure that the model was truly doing what they wanted. The engineers had nailed it!

When they finally built a production unit and put it on-line, the customer was so happy with the results that he refused to take lenses or reflectors that were not made on a line having this rim check inspection equipment. Of course, the solution was to build additional units for the other production lines, which was done!

Innovation is not always easy, nor does it always go on a smooth path. For example, the engineers had to build the electronic mask so that the device would ignore flashes coming from anywhere other than the outside rim of the glass. That is why management often doesn't like innovation—there are always risks and some additional delays or costs. However, with good project overview and the use of Six Sigma tools such as the QFD and FMEA, cost overruns and delays are minimized.

Webster's *New World Dictionary*'s definition for serendipity is "an apparent aptitude for making fortunate discoveries accidentally." Innovators have that aptitude, and it was in play when the engineer picked up the laser pointer and shone it parallel to the rim of the glass containing a rim check, causing the check to glow! He knew at that point that he had "accidentally" discovered the secret to finding rim checks on-line! It was his innovative spirit that caused him to even try to shine the laser on the rim of the glass to see if it would assist in being able to see the rim check.

In Chapter 3 and Chapter 15, a case study on a test cutting machine was briefly discussed. Here is more detail on that project that shows the degree of innovation involved.

CASE STUDY: MAKING A CUTTING MACHINE ACTUALLY CUT GLASS TUBING

This glass tubing plant had 12 production lines, and each line had a tubing cutter that was to cut lengths of tubing from a continuous moving glass tubing line. This was the most Rube Goldberg–looking device you can imagine. It certainly did not look like a glass tubing cutter. It had a continuously rotating arm with a spring-mounted piece of U-shaped aluminum hanging loosely on its end. The circumference of the circle of this rotating arm matched the desired tube length to be cut from the continuously moving glass tubing line. Ideally, each time the piece of U-shaped aluminum hit the moving glass tubing below, the aluminum "cutter" would match the speed of the tubing line and make a cut such that an individual length of tubing came off the line.

However, in all the times the engineer had visited this glass plant, he never saw this supposed cutting device really work. It would smash the tubing at regular intervals, so there was broken glass all over the place. Not only was this dangerous and causing a lot of downtime because of wrecks, but because of the poor quality of the cut, a great deal of extra tubing had to be included in the cut length to allow for a second off-line finished cut whose original purpose was just to trim up the ends.

This problem had been going on for many years, and it was such a nuisance that alternative (and much more complex and expensive) cutter designs had been tried. But these complicated devices were a nightmare to maintain and could not keep up with the high-speed lines. There also were many studies done on trying to understand the current cutting process sufficiently to make it work. One such recent study took a glass technologist with his doctorate in glass physics a year. At the conclusion of his study, he went to the plant with some modified equipment and asked the plant to try the equipment changes and the process as he envisioned it. He was not successful.

Here is some more detail on the "cutting" device. As mentioned, at the end of the rotating arm was a U-shaped piece of aluminum that swung freely. As this aluminum piece went around with the rotating arm, at the top of the arm swing, the aluminum piece rubbed against a wet sponge. This kept the U-shaped aluminum piece wet and cool. As the aluminum piece came down and around and then touched the moving glass tubing, it traveled and stayed in contact with the hot tubing for a few tenths of a second before it cut (or smashed) the tubing.

Few people had ever seen this device actually cut the glass. However, several operators claimed that periodically, for very short intervals of time, the glass would break off very cleanly. The cutter would run that way for a minute or two before going back to being a glass smasher!

Since this device was apparently capable of running well, an innovative engineer decided that he was going to try to capture whatever process variables were needed to have this device work well all the time. Then he would give the operators whatever tools, training, or modified equipment they needed to run the cutter correctly.

Before doing anything, the engineer reviewed all the past studies that had been done on this device, including the most recent one-year study that had just ended. The physicist who had done this recent study explained to the engineer what he thought were the physics going on that would enable this cutter to work. But this physicist was unable to turn his theory into practice. The engineer then talked to several of the best operators, asking them what they had seen when the device was working. They all said that the cutter made a very clear ringing sound, a "ping," when it was cutting the tubes successfully. The operators could identify no specific changes in the process between the times when the cutter worked and the times when it did not work.

The engineer was very active in implementing Six Sigma, and one of the things you do in Six Sigma is to have a meeting with everyone who is involved in a project and do a QFD to get their inputs. The engineer called such a meeting on this project; but most of the things that were mentioned in the meeting had already been thought of. However, the two operators who attended and who had actually seen this device working came up with something that none of the rest of the attendees had thought of. The operators were adamant that the tubing support wheel under the glass at the cutting point should be spring-mounted. Since the rotating aluminum cutter piece that hit the top of the tubing was already spring-mounted, most of the people in the meeting thought that another spring on the support wheel would be redundant. However, since the operators were insistent, the innovative engineer decided to add the support wheel spring to the QFD action list. He felt that if the additional spring wasn't of any use, he could just put in a very stiff spring that would make it ineffective. Some of the other engineers in the meeting were very unhappy about the inclusion of this spring, since the operators could not explain the need for it. But the engineer in charge of the project had had earlier experience in seeing how hourly process employees sometimes had feelings about a process need that they could not quantify or detail with logic. And these two operators were the only ones in the room that had actually seen this device work!

The next step was to build a cutter test device that had adjustments, pressure sensors, and indicators in all identified locations so that the engineers could experiment and find what settings were critical. If there was a process, and the operators insisted that there was, the goal was to capture it.

Months later, once the test device had been built and brought to the production plant, a group of engineers was there to try to find this mystical cutting process. They were going to run a modified design of experiments in which many different combinations were going to be tried. There was to be a lot of judgment going on during the tests based on real-time observations, because there wasn't enough time to run all the theoretical test combinations. The engineers were going to chase any hint of a process, even if it wasn't on the initial trial list.

The test device was put in line, and the engineers started adjusting different variables. Eventually it came time to try the operator-suggested spring on the support wheel, but it turned out that the spring that had been chosen was too weak to support the wheel. While the plant people went to find a somewhat stronger spring, the engineer in charge of the project stepped up to support the wheel so that the engineers could continue with their tests. Without this, the tubing would not have been cut and would have gone down a scrap hole to return to the furnace. Manually supporting this wheel wasn't without some risk because the moving glass tubing at that position was quite dangerous. So for safety, the engineer put on protective gloves that went to his shoulders before reaching in to support the wheel.

While he was holding up this support wheel, there were all kinds of smart-mouth comments, like the engineer now being a critical part of the process. One thing the innovative engineer believed about this kind of test was that people should pay attention rather than making smart remarks, because it is one of the rare times when things are running with a lot of controls. Every eye, ear, nose, and brain should be focused. In fact, he quickly called to everyone's attention that if they hadn't been so busy making smart remarks, they would have noticed that he now had a perfect cut, with the glass tubes ringing off exactly as the operators had told the engineers they would! The "ping" that the operators described was apparent. Immediately everyone started taking readings on every adjustment so that they could capture the key process variables. It appeared that if the support wheel was just lightly supported with the minimum of upward force, then all the other variables had a fairly wide operating window.

Further tests were run to determine the optimum settings for all the other variables, along with their acceptable ranges. All this information would be needed to build the production units.

The question was, how did the operators ever see this cutting process work correctly when there was no spring on the support wheel before? The engineer's guess was that at times, the glass tubing bowed slightly upward off the support wheel, so that the tubing itself was the required spring. But that was just the engineer's theory. No one ever really knew!

Now that they had seen what was required for a good cut, the engineers could go back and try to understand the physics involved, which were quite complex. They involved both mechanical and thermal forces.

Whoever had originally invented this clever device must have understood what was going on, but he had neglected to pass on his knowledge, and all his understanding had been lost. Maybe the support wheel was originally made of a more pliant material that didn't require a support spring. No one knew. There is no need to go into the detailed physics here, but the engineers felt that they now understood the cutting process and how it worked.

Over the following year, production versions of this unit were built and installed on all the production lines. These devices saved well over $1.5 million per year by eliminating the extra glass that had had to be added to the tube length to make the finished cut successful. Using Six Sigma data analysis, the engineers could see that the cutter went from essentially zero effective cuts to 99.9 percent performance on acceptable cuts. The new design also eliminated the safety hazard of broken glass.

Now that the process was understood, the engineers were able to apply the same techniques on small tubing in another plant. In several cases, when the tubing length was very short, multiple aluminum cutters spaced an equal distance around the rotating arm were required, which was another innovation detail.

A side note on the device that the engineer put into another plant to cut small tubing: that plant had never had a cutter like this, and no one was sure that it would even work on small-diameter tubing. The usual Six Sigma QFD and FMEA were done before installing this device, and several unique needs of the plant were identified and incorporated. The unit was put on-line in the morning, and by noon it was cutting small tubing quite well. Therefore, the engineers went to lunch, with the intent of coming back after lunch and experimenting with different settings to see how sensitive they all were and to find the optimum settings.

By the time the engineers came back from lunch, the new cutter was off-line. The engineers tried to find out why, and all they got were angry answers from the production people. They said that right after the engineers had left the plant, the plant chief engineer had come and had told the maintenance people to take the cutter out of the production line. No one had seen any issues, so no one knew why he had done this. The chief engineer had gone to lunch, and he had left no note. The engineer in charge of the cutter installation went to the plant manager and told him what had happened, including the fact that the chief engineer had refused to attend the Six Sigma QFD and FMEA that were done on the project before it was built and installed.

Several weeks later, the plant chief engineer was taken out of his job. Apparently this chief engineer acted in a similar fashion anytime anything new was brought into his plant. The plant then put the tubing cutter back on line, and everyone (except perhaps for the former chief engineer) was happy with the cutter.

In this project, the line operators gave strong inputs to the innovation requirements by insisting on the spring on the support wheel. And this need was included on the Six Sigma QFD priority list. As noted several times before, innovation often comes from people who perhaps don't know the technical reasons but have a sense of feel on what is important.

The barrier on this project was that no one understood the underlying physics that made the cutter work. Others had tried to solve this problem without having that full understanding. The solution was to build an instrumented test cutter that included all the key variables, including the one suggested by the operators, so that the engineers could capture and then understand the physics of the cutting process.

There is another side story with this project. Just after the engineer got formal approval to make an instrumented unit to capture the process, an "Edison Engineer" joined the engineer's group for six months. The GE Edison Engineering Development Program is for very bright engineering graduates, generally from elite engineering schools, who go on three six-month assignments with various engineers or managers. These Edison Engineers are such prima donnas that many engineers and managers don't want to work with them, and if an Edison Engineer doesn't succeed in her assignment, it is assumed that the training engineer or manager screwed up. But this engineer was confident of his own abilities, so he liked Edison Engineers. He gave the Edison Engineer the assignment of coordinating the purchase and then the installation of this instrumented tubing cutter into the plant.

At the beginning of the Edison Engineer's assignment, the engineer was reviewing this project with her. She suddenly had a very frightened look on her face. She said, "You are not sure this is going to work, are you?" The engineer in charge of this project then realized that this very bright Edison Engineer, who had apparently never gotten anything but A's in school, had never worked on anything that had any risk such that hard work alone would not guarantee success! The engineer told the Edison Engineer that they would probably learn something, but that indeed he could not assure her of some breakthrough. She accepted this, but was obviously not very happy.

Once the device was on-line, cutting tubes beautifully with the critical setting ranges identified, the Edison Engineer came up to the engineer in charge. She was obviously ecstatic, but she also said that the innovative engineer had thrown out everything she had learned in four years of college. He asked her why she felt that way. She said, "Well, on the QFD, you included the spring below the support wheel, which no one could logically justify and the other engineers did not want. And that turned out to be the most important factor. And you supported the tubing by hand while they were looking for the proper spring, and that somewhat dangerous act enabled you to see that minimal force holding up the wheel was critical. And you did not follow the design of experiments plan that was set up before the tests started." The team engineer's answer was that his job as her training engineer, as he saw it, was to give her some practical knowledge to go along with the largely theoretical things that she had learned in college, as both are critical for success in any job.

He explained to her that in his years of working with hourly employees, he had often seen them come up with insights that they could not explain logically. For example, in this case, it could have been the tubing bowing up slightly that was causing the spring effect, and it could have appeared to the operators as if the support wheel itself was having some "spring." As for supporting the wheel manually, giving the insight that minimum support on the wheel was important, his experience was that serendipity often comes to those who keep trying and use all their skills and senses on a problem. And since he was using the operator's protective gloves, he did not feel that what he did was dangerous at all. As for not following the complete design of experiments plan, as you are testing a device you are constantly getting smarter, so you should feel free to modify your design plan accordingly. He told her that he couldn't even imagine not doing that!

Edison Engineers give their six-month assignment presentations in front of each other, the training engineer, their respective managers and general managers, and the vice president of engineering. The Edison Engineer was very nervous and was really working hard preparing for her presentation. When she reviewed it with the engineer in charge, he laughed and told her that she could mess up her whole presentation; but when she got to the end and showed how this project was going to save GE more than $1.5 million a year, that was all that the vice president would hear. The other Edison Engineers would have annual savings of perhaps $10,000 per year.

This is exactly what happened in her presentation. She stole the show. When the vice president talked to her after her presentation, he told her that the only bad thing about her project was that she might not have such a successful project again in her life, and the vice president did not want her to think that every project always comes out that well!

This project was very rewarding for everyone involved, including the hourly employees who insisted on putting in the spring below the support wheel. The engineer in charge made sure that the plant manager and the other employees at the plant all knew how the operators' inputs had made this project a success.

This project illustrated almost all the traits that are common among manufacturing innovators. The engineer certainly had a high enough level of confidence to be willing to take on a project that had defeated many people before him. He questioned why others had not listened to the operators' observations that at times the cutter worked well. Because of this, he surmised that there was some available combination that would enable the cutter to work without designing a totally new cutter. He was sensitive to the input from the hourly workers that the additional spring be added. His persistence was such that he did not even want to wait while the production people went for a stronger spring, and this impatience led to him to discover that minimal wheel support was critical. And while he was holding up the wheel, he was still focused on the goal, so that he saw what others were missing—that he was getting an excellent cut. The engineer did not need to pursue esoteric solutions; a simple added spring on the current cutter sufficed! Once the engineer saw the breakthrough that the spring gave, he abandoned the earlier design of experiments plan and immediately started finding the optimum settings of the other adjustments to support this breakthrough finding.

WHAT WE HAVE LEARNED IN CHAPTER 20

Besides general engineering skills, successful manufacturing innovators have:

1. A high sense of confidence bordering on ego
2. Little fear of authority—everything is questioned
3. The ability to work around or through barriers
4. Persistence to the point that serendipity starts to be evident
5. An insistence on excellence

6. The ability to work with hourly people—to get their trust and build on their intimate observations of the process
7. The ability to "see" a problem and its related opportunity with more focus and clarity than most people
8. The ability to recognize when available solutions can apply to unique areas.

The Six Sigma Statistical Tool Finder Matrix

Look in the Need and Detail columns in the matrix given here to identify the appropriate Six Sigma tool.

Need	Detail	Six Sigma Tool	Location
Address customer needs by identifying and prioritizing actions	Convert qualitative customer input to specific prioritized actions	Simplified QFD	Chapter 3
Minimize collateral damage caused by a product or process change	Convert qualitative input on concerns to specific prioritized actions	Simplified FMEA	Chapter 4
Identify key process input variables (KPIVs)	Use expert input on a process	Cause-and-effect fishbone	Chapter 5
	Use historical data	Correlation tests	Chapter 7
Pinpoint possible problem areas of a process	Use expert input	Process flow diagram	Chapter 6
Verify measurement accuracy	Determine whether a gauge is adequate for the task	Simplified gauge verification	Chapter 9

Need	Detail	Six Sigma Tool	Location
Calculate minimum sample size	Variables (decimal) data	Sample size—variables	Chapter 12
	Proportional data	Sample size—proportions	Chapter 13
Determine whether there is any statistically significant change in variables (decimal) data	Compare a sample with historical (population) data	1. Plot data	Chapter 11
		2. Chi-squared test	Chapter 12
		3. t test	Chapter 12
	Compare two samples to each other	1. Plot data	Chapter 11
		2. F test	Chapter 12
		3. t test	Chapter 12
Determine whether there is any statistically significant change in proportional data	The mathematical probability of the population is known	Excel's BINOM.DIST	Chapter 10
	Compare a sample with a population where both proportions are calculated	Sample/population formula	Chapter 13
	Compare two samples where both proportions are calculated	Sample/sample formula	Chapter 13
Minimize process defect excursions	Applicable to variables (decimal) data	Simplified control charts	Chapter 16
Optimize key process input variable settings (KPIVs)	Use on a current process	Simplified DOE	Chapter 15
Determine whether tolerances are appropriate	How were tolerances determined?	Need-based tolerances	Chapter 17
	Are any stacked parts with tolerances at worst case?	RSS tolerances	Chapter 17
Identify variation contributed to the process or assembly variation by each component	Identifies problem components	Simplified transfer function	Chapter 18

APPENDIX B

Six Sigma Tool Checkoff List

The example of the Six Sigma tool checkoff list given here can be used as a guideline for building a tool checkoff list specific to your project. This example was used for a project to develop a laser and digital camera inspection device to measure the rim diameter of a product (case study in Chapter 2). Initial tests had shown that a series of lasers, if directed at the rim of the product, could project an image onto digital cameras, enabling a quick inspection of the rim diameter. The purpose of the project was to replace a slower gauge that used dial indicators. Although the dial indicator gauge gave satisfactory measurements, it was too slow to keep up with a planned increase in line speed.

Note that in this Six Sigma tool checkoff list, many of the tools are used in more than one part of the DMAIC process. For example, simplified gauge verification is used in the Measure, Improve, and Control steps, and the tests for significant change are used in the Analyze, Improve, and Control steps, sometimes comparing two samples and sometimes comparing the population to a sample. Where and how often each tool will be used is dictated by the individual project. Some tools, like the QFD and FMEA, should be used on *every* project, since input from people is always required.

This Six Sigma tool checkoff list is sometimes modified during a program. As issues arise, the proposed solutions often include the use of additional Six Sigma tools. Each modification of the Six Sigma tool checkoff list should be dated, and prior lists should be kept for reference as part of the program file. The use of the Six Sigma tool checkoff list should be in addition to whatever total program management tool is used. It is not a replacement for traditional program management techniques.

DMAIC	Tool	Use
Define	Simplified QFD	Get input on the detailed needs of the project from inspectors, line maintenance, quality engineers, design engineers, production manager, and customers. Develop action list.
	Simplified FMEA	Identify any other processes that may be negatively affected by implementing this project. (For example, the extra time required of line maintenance people may negatively affect the maintenance of other equipment.) Develop action list.
	Fishbone diagram	Identify all the input variables causing the diameter to vary, to make sure that we have samples of all types of diameter variations. This enables us to verify that all variations of diameters can be measured using the laser and digital camera device.
Measure	Calculate minimum sample size for variables data	Determine the number of sample products required to run statistically valid tests on the new equipment.
	Simplified gauge verification	Using product masters, verify that the *test setup* can measure the product diameter sufficiently accurately.
Analyze	Test for significant change with variables data (decimals) two-sample test	Run tests on the sample products using both the traditional manual dial indicator measurement method and the test setup. Verify that there is no significant difference between the two methods. If a difference is found, additional tests will be required to identify the reason(s).
Improve	Simplified gauge verification	Using product masters, verify that the *production device* can measure the product diameter sufficiently accurately.
	Test for significant change with variables data (decimals) two-sample test	Run tests on the product samples using both the traditional manual measurement method and the *production device*. Verify that there is no significant difference between the two methods. If a difference is found, additional tests will be required to identify the reason(s).

DMAIC	Tool	Use
Improve	Test for significant change with variables data (decimals) population/sample	Run tests on the production products using the new device. Verify that there is no significant difference between the new data and historical (population) data. If a difference is found, additional investigation will be required to determine why.
Control	Test for significant change with variables data (decimals) population/sample	Using the output data, continuously calculate results on a running sample and compare the results with historical data. If the running sample results are statistically different from the historical population results, then the device may not be working properly. Line maintenance would then be alerted.
	Simplified control charts	Using the measurement output of the new device, implement a simplified control chart to help the operator reduce diameter excursions.
	Simplified gauge verification	Using product masters, verify that the production device can measure the product diameter sufficiently accurately. Schedule this simplified gauge verification monthly.

Formulas Used
in This Book

From Chapter 9

Simplified Gauge Verification, Variables Data

$$\% \text{ gauge error} = \frac{5s + 2 \times |\text{master} - \bar{x}|}{\text{tolerance}} \times 100 \text{ (ideally} <10\%, \text{ maximum } 30\%)$$

$\bar{x}$ = average of *all* 21 readings of a master product

s = standard deviation of *all* 21 readings of the master product

master = standardized dimension of the master product

tolerance = allowable product tolerance (max – min)

$|\text{master} - \bar{x}|$ = difference between master and average $\bar{x}$, ignoring minus signs

Use on Problem Type

You wish to verify that a gauge that is being used to measure variables data is not "using up" more than 30 percent of the tolerance. Have three inspectors read a prequalified master product seven times each, getting 21 total readings. The average and standard deviation for these data are calculated and entered into the formula given here.

Both repeatability/reproducibility (the ability to duplicate a reading) and accuracy (the aim, or correctness of the average reading) are included in the simplified gauge verification formula. The accuracy portion is:

$$2 \times |\text{master} - \bar{x}|$$

From Chapter 12

Estimating Population $\bar{X}$ and S from Multiple Samples of Similar Size n, Variables Data

$$\bar{X} = \frac{\bar{x}_1 + \bar{x}_2}{2}$$

$\bar{X}$ is the population average

$\bar{x}_1$ is the average from the first sample

$\bar{x}_2$ is the average from the second sample

$$S = \sqrt{\frac{s_1^2 + s_2^2}{2}}$$

S is the population sigma

s_1 is the sigma of the first sample

s_2 is the sigma of the second sample

If you have three or more samples, modify the formulas accordingly, with more sample sigma or averages in the numerator and dividing by the total number of samples.

Use on Problem Type

You have multiple samples of variables data with similar sample sizes, and you want to estimate the population's average and standard deviation.

From Chapter 12

Calculating Minimum Sample Size and Sensitivity, Variables Data

$n = \left(\dfrac{Z \times S}{h}\right)^2$ to calculate minimum sample size on variables data

n = minimum sample size on variables data (always round up)

Z = confidence level (when in doubt, use $Z = 1.96$)

S = the population standard deviation

h = the smallest change we want to be able to sense
(When in doubt, use h = total tolerance/10, or $h = 0.6S$.)
   Note that the formula just given can be rewritten as:

$$h = \sqrt{\dfrac{Z^2 \times S^2}{n}}$$

   This allows us to see what sensitivity h (change) we can expect to see with a given sample size and confidence level.

Use on Problem Type

Use this formula to calculate the minimum variables data sample size you need to make estimates on a population. To use this formula, you have to define what change, or sensitivity h, you want to be able to sense. A rule of thumb is to use the total tolerance/10.

   By rewriting the formula, you can solve directly for the change, or sensitivity h. This provision is used when the sample size has already been determined and you want to see what change, or sensitivity h, you will be able to sense.

From Chapter 12

Chi-Squared Test Value of a Sample Sigma s Versus a Population Sigma S, Variables Data

$$\text{Chi}_t^2 = \frac{(n-1)s^2}{S^2}$$

n = sample size

s = sample sigma

S = population sigma

We compare the calculated Chi_t^2 results with the values in the simplified chi-squared distribution table (Exhibit 12-1). If the Chi_t^2 test value we calculated is less than the table low value or greater than the table high value, we are 95 percent confident that the sigma s of the sample is different from the sigma S of the population.

Use on Problem Type

This formula is used with variables data when we want to test whether a sample's sigma is significantly different from the population's sigma. We compare the calculated result from this formula with the values in the simplified chi-squared distribution table (Exhibit 12-1) to determine whether there is a statistically significant difference between the two.

This formula is normally used after we have already compared the plots of data from both the sample and the population and are satisfied that the shapes of the two distributions of data are not dramatically different.

From Chapter 12

t Test of a Population Average $\overline{X}$ Versus a Sample Average $\overline{x}$, Variables Data

$$t_t = \frac{|\overline{x} - \overline{X}|}{\dfrac{s}{\sqrt{n}}}$$

$\overline{X}$ = population average

$\overline{x}$ = sample average

s = sample sigma

n = sample size

$|\overline{x} - \overline{X}|$ is the absolute value of the difference of the averages, so ignore a minus sign in the difference.

We then compare this calculated *t*-test (t_t) value against the value in the simplified *t*-distribution table (Exhibit 12-2). If our calculated *t*-test (t_t) value is greater than the value in the table, then we are 95 percent confident that the sample average is significantly different from the population average.

Use on Problem Type

Use this formula on variables data when you want to test whether a sample's average is significantly different from a population's average. We compare the calculated result from this formula with the values in the simplified *t*-distribution table (Exhibit 12-2) to determine whether there is a statistically significant difference between the two.

This formula is normally used after we have already compared the plots of data from both the sample and the population and are satisfied that the shapes of the two distributions of data are not dramatically different and after the chi-squared test of sigma did not show a significant difference.

From Chapter 12

F Test Comparing Two Samples' Sigma *s*, Variables Data

$$F_1 = \frac{s_1^2}{s_2^2} \text{ (put the larger } s \text{ on top, in the numerator)}$$

s_1 = sample with the larger sigma

s_2 = sample with the smaller sigma

The sample sizes n should be within 20 percent of each other. There are tables and programs that allow for greater differences, but since you can control sample sizes, and since you get more reliable results with similar sample sizes, these other tables and programs are generally not needed.

Compare this F_t with the value in the simplified *F* table (Exhibit 12-3). If the F_t value exceeds the table *F* value, then the sigma are significantly different.

Use on Problem Type

This formula is used on variables data when we want to test whether two samples' sigma are significantly different. We compare the calculated result from this formula with the values in the simplified *F* table (Exhibit 12-3) to determine whether there is a statistically significant difference between the two.

This formula is normally used after we have already compared the plots of data from both of the samples and are satisfied that the shapes of the two distributions of data are not dramatically different.

From Chapter 12

t Test of Two Sample Averages $\bar{x}_1$ and $\bar{x}_2$, Variables Data

$$t_t = \frac{\left|\bar{x}_1 - \bar{x}_2\right|}{\sqrt{\left(\dfrac{n_1 s_1^2 + n_2 s_2^2}{n_1 + n_2}\right)\left(\dfrac{1}{n_1} + \dfrac{1}{n_2}\right)}}$$

$\bar{x}_1$ and $\bar{x}_2$ are two sample averages

s_1 and s_2 are the sigma on the two samples

n_1 and n_2 are the two sample sizes

$\left|\bar{x}_1 - \bar{x}_2\right|$ is the absolute difference between the averages, ignoring a minus sign in the difference.

We then compare this calculated *t*-test value against the value in the simplified *t*-distribution table (Exhibit 12-2). If our calculated *t*-test number is greater than the value in the table, then we are 95 percent confident that the sample averages are significantly different.

Use on Problem Type

Use this formula on variables data when you want to test whether the averages of two samples are significantly different. We compare the calculated result from this formula with the values in the simplified *t*-distribution table (Exhibit 12-2) to determine whether there is a statistically significant difference between the two.

This formula is normally used after we have already compared the plots of data from the two samples and are satisfied that the shapes of the two distributions of data are not dramatically different and after an *F* test of sigma did not show a significant difference.

From Chapter 13

Calculating Minimum Sample Size and Sensitivity, Proportional Data

$$n = \left(\frac{1.96\sqrt{(p)(1-p)}}{h}\right)^2$$

n = sample size of attribute data, like "good" or "bad" (95 percent confidence)

p = probability of an event (the proportion of defects in a sample, chance of getting elected, or some similar classification) (When in doubt use $p = 0.5$, the most conservative.)

h = sensitivity, or accuracy required

(For example, for predicting elections, the sensitivity required may be ±3 percent, or $h = 0.03$. Another guideline is to be able to sense 10 percent of the tolerance or difference between the proportions.)

Note that the formula shown here can be rewritten as:

$$h = 1.96\sqrt{\frac{(p)(1-p)}{n}}$$

This allows us to see what sensitivity h we will be able to achieve with a given sample size and probability.

Use on Problem Type

Use this formula to calculate the minimum proportional data sample size you need in order to make estimates on a population with 95 percent confidence. To use this formula, you have to define what change, or sensitivity h, you want to be able to sense. A rule of thumb is to use the total tolerance/10.

By rewriting the formula, you can solve directly for the change, or sensitivity h. This provision is used when the sample size has already been determined and you want to see what change, or sensitivity h, you will be able to sense.

From Chapter 13

Comparing a Proportional Sample with the Population (95 Percent Confidence)

First, we must calculate a test value Z_t.

$$Z_t = \frac{|p - P|}{\sqrt{\dfrac{P(1 - P)}{n}}}$$

P = proportion of defects (or whatever) in the population (historical)

p = proportion of defects (or same as above) in the sample

$|p - P|$ = absolute proportion difference (no minus sign in difference)

n = sample size

If $Z_t > 1.96$, then we can say with 95 percent confidence that the sample is different from the population.

Use on Problem Type

Use this formula with proportional data when you want to test whether a sample is significantly different from a population.

From Chapter 13

Comparing Two Proportional Data Samples (95 Percent Confidence)

We must calculate a test value Z_t.

$$Z_t = \frac{\left|\dfrac{x_1}{n_1} - \dfrac{x_2}{n_2}\right|}{\sqrt{\left(\dfrac{x_1 + x_2}{n_1 + n_2}\right)\left(1 - \dfrac{x_1 + x_2}{n_1 + n_2}\right)\left(\dfrac{1}{n_1} + \dfrac{1}{n_2}\right)}}$$

x_1 = number of defects (or whatever) in sample 1

x_2 = number of defects (or same as above) in sample 2

$\left|\dfrac{x_1}{n_1} - \dfrac{x_2}{n_2}\right|$ = absolute proportion difference (no minus sign in difference)

n_1 = size of sample 1

n_2 = size of sample 2

If $Z_t > 1.96$, then we can say with 95 percent confidence that the two samples are significantly different.

Use on Problem Type

Use this formula with proportional data when you want to test whether two samples are significantly different from each other.

From Chapter 17

RSS: Calculating the Sigma *S* from Multiple Parts Stack-Up

$$S = \sqrt{(1.3s_1)^2 + (1.3s_2)^2 + (1.3s_3)^2 + \cdots}$$

S = the resultant assembly stack-up sigma

$s_1, s_2, s_3, \ldots$ = the sigma of each individual part being stacked

Each sigma s is multiplied times 1.3 to allow for a long-term sigma drift. If each of n stacked-up parts has the same sigma s, then:

$$S = \sqrt{n(1.3s)^2}$$

Use on Problem Type

If multiple parts are stacked in an assembly, the tolerances on those parts are likely to have been calculated using worst-case methods. Using the RSS method of calculating tolerances can open tolerances on those parts or show that the assembly has less variation than assumed.

From Chapter 18

RSS Linear Transfer Functions

$$S_t = \sqrt{s_1^2 + s_2^2 + s_3^2 + s_4^2 \cdots}$$

S_t = the critical sigma of the total assembly or process

$s_1, s_2, s_3, s_4, \ldots$ are the sigma of the variables linearly affecting the critical sigma of an assembly or process

Each variable's influence must be stated in common units that are consistent with the part, assembly, or process. For example, if we are studying the thickness variation of an injection-molded part and one of the contributing variables is the weight of the injected raw material, we need to state that variable's sigma in "thickness variation per sigma," rather than in "weight unit per sigma."

Use on Problem Type

Use the simplified linear transfer function to understand the effect of each component on the total variation of a part, an assembly, or a process. The sum of the squares of the contributing variables' sigma must equal the square of the sigma of the total assembly or process. If the sum is too low, one or more variables are missing. Each sigma contribution must have units that are consistent with the product effect being measured. If this is the case, it is valid to compare the sigmas to see which variable is most critical.

This formula is not applicable to nonlinear processes, such as many chemical processes that have complex interactions among input variables and therefore require nonlinear transfer functions. Nonlinear transfer functions, which require partial derivatives, are beyond the scope of this book (and most Six Sigma work).

From Chapter 19

Estimating the Parent Population Sigma *S* from the $s_{\bar{x}}$ of a Child Distribution

$$S = s_{\bar{x}} \sqrt{n}$$

S = parent population sigma (sigma based on the raw data)

$s_{\bar{x}}$ = child distribution sigma (sigma of the multiple sample averages)

n = individual sample size (quantity in each raw data sample)

Use on Problem Type

This formula is needed in order to compare the data collected by many quality departments with samples taken in the course of Six Sigma work. Quality department systems often use child distributions, which are based on multiple sample averages, rather than the raw data themselves. The raw data are often discarded after the average is calculated.

APPENDIX D

Tables Used
in This Book

Abbreviated Binomial Table

Values within the table are the probability P of getting exactly x successes in n trials.

n # of trials	x successes on n trials	p (each trial) =	0.125 (1/8)	0.167 (1/6)	0.250 (1/4)	0.500 (1/2)
2	0		0.7656	0.6944	0.5625	0.2500
2	1		0.2188	0.2778	0.3750	0.5000
2	2		0.0156	0.0278	0.0625	0.2500
		Sum of P:	1.0000	1.0000	1.0000	1.0000
3	0		0.6699	0.5787	0.4219	0.1250
3	1		0.2871	0.3472	0.4219	0.3750
3	2		0.0410	0.0694	0.1406	0.3750
3	3		0.0020	0.0046	0.0156	0.1250
		Sum of P:	1.0000	1.0000	1.0000	1.0000
4	0		0.5862	0.4823	0.3164	0.0625
4	1		0.3350	0.3858	0.4219	0.2500
4	2		0.0718	0.1157	0.2109	0.3750
4	3		0.0068	0.0154	0.0469	0.2500
4	4		0.0002	0.0008	0.0039	0.0625
		Sum of P:	1.0000	1.0000	1.0000	1.0000
5	0		0.5129	0.4019	0.2373	0.0313
5	1		0.3664	0.4019	0.3955	0.1563
5	2		0.1047	0.1608	0.2637	0.3125
5	3		0.0150	0.0322	0.0879	0.3125
5	4		0.0011	0.0032	0.0146	0.1563
5	5		0.0000	0.0001	0.0010	0.0313
		Sum of P:	1.0000	1.0000	1.0000	1.0000
10	0		0.2631	0.1615	0.0563	0.0010
10	1		0.3758	0.3230	0.1877	0.0098
10	2		0.2416	0.2907	0.2816	0.0439
10	3		0.0920	0.1550	0.2503	0.1172
10	4		0.0230	0.0543	0.1460	0.2051
10	5		0.0039	0.0130	0.0584	0.2461
10	6		0.0005	0.0022	0.0162	0.2051
10	7		0.0000	0.0002	0.0031	0.1172
10	8		0.0000	0.0000	0.0004	0.0439
10	9		0.0000	0.0000	0.0000	0.0098
10	10		0.0000	0.0000	0.0000	0.0010
		Sum of P:	1.0000	1.0000	1.0000	1.0000

Standardized Normal Distribution Table

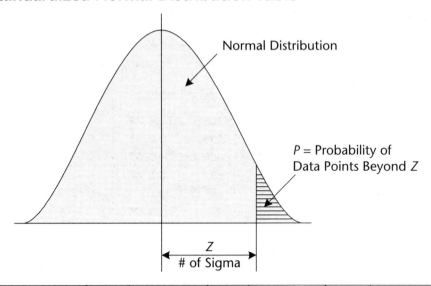

Z	P	Z	P	Z	P	Z	P
0.00	0.5000	0.05	0.4801	0.10	0.4602	0.15	0.4404
0.20	0.4207	0.25	0.4013	0.30	0.3821	0.35	0.3632
0.40	0.3446	0.45	0.3264	0.50	0.3085	0.55	0.2912
0.60	0.2743	0.65	0.2578	0.70	0.2420	0.75	0.2266
0.80	0.2119	0.85	0.1977	0.90	0.1841	0.95	0.1711
1.00	0.1587	1.05	0.1469	1.10	0.1357	1.15	0.1251
1.20	0.1151	1.25	0.1056	1.30	0.09680	1.35	0.08851
1.40	0.08076	1.45	0.07353	1.50	0.06681	1.55	0.06057
1.60	0.05480	1.65	0.04947	1.70	0.04457	1.75	0.04006
1.80	0.03593	1.85	0.03216	1.90	0.02872	1.95	0.02559
2.00	0.02275	2.05	0.02018	2.10	0.01786	2.15	0.01578
2.20	0.01390	2.25	0.01222	2.30	0.01072	2.35	0.009387
2.40	0.08198	2.45	0.007143	2.50	0.006210	2.55	0.005386
2.60	0.004661	2.65	0.004025	2.70	0.003467	2.75	0.002980
2.80	0.002555	2.85	0.002186	2.90	0.001866	2.95	0.001589
3.00	0.001350	3.05	0.001144	3.10	0.0009677	3.15	0.0008164
3.20	0.0006872	3.25	0.0005771	3.30	0.0004835	3.35	0.0004041
3.40	0.0003370	3.45	0.0002803	3.50	0.0002327	3.55	0.0001927
3.60	0.0001591	3.65	0.0001312	3.70	0.0001078	3.75	0.00008844
3.80	0.00007237	3.85	0.00005908	3.90	0.00004812	3.95	0.00003909
4.00	0.00003169	4.05	0.00002562	4.10	0.00002067	4.15	0.00001663
4.20	0.00001335	4.25	0.00001070	4.30	0.00000855	4.35	0.00000681
4.40	0.00000542	4.45	0.00000430	4.50	0.00000340	4.55	0.00000268
4.60	0.00000211	4.65	0.00000166	4.70	0.00000130	4.75	0.00000102
4.80	0.00000079	4.85	0.00000062	4.90	0.00000048	4.95	0.00000037

Simplified Chi-Squared Distribution Table

To test a sample sigma s (with sample size n) vs. a population of sigma S

	95% Confident They Are Different if Chi_t^2 Is				95% Confident They Are Different if Chi_t^2 Is		
	< Low	Or	> High		< Low	Or	> High
n	Low Test		High Test	n	Low Test		High Test
6	0.831209		12.83249	36	20.56938		53.20331
7	1.237342		14.44935	37	21.33587		54.43726
8	1.689864		16.01277	38	22.10562		55.66798
9	2.179725		17.53454	39	22.87849		56.89549
10	2.700389		19.02278	40	23.65430		58.12005
11	3.246963		20.48320	41	24.43306		59.34168
12	3.815742		21.92002	42	25.21452		60.56055
13	4.403778		23.33666	43	25.99866		61.77672
14	5.008738		24.73558	44	26.78537		62.99031
15	5.628724		26.11893	45	27.57454		64.20141
16	6.262123		27.48836	46	28.36618		65.41013
17	6.907664		28.84532	47	29.16002		66.61647
18	7.564179		30.19098	48	29.95616		67.82064
19	8.230737		31.52641	49	30.75450		69.02257
20	8.906514		32.85234	50	31.55493		70.22236
21	9.590772		34.16958	55	35.58633		76.19206
22	10.28291		35.47886				
23	10.98233		36.78068	60	39.66185		82.11737
24	11.68853		38.07561	65	43.77594		88.00398
25	12.40115		39.36406				
26	13.11971		40.64650	70	47.92412		93.85648
27	13.84388		41.92314	80	56.30887		105.4727
28	14.57337		43.19452				
29	15.30785		44.46079	90	64.79339		116.989
30	16.04705		45.72228	100	73.3611		128.4219
31	16.79076		46.97922				
32	17.53872		48.23192				
33	18.29079		49.48044				
34	19.04666		50.72510				
35	19.80624		51.96602				

Simplified *t*-Distribution Table

To compare a sample average (size = *n*) with a population average or to compare two samples of size n_1 and n_2, using $n = n_1 + n_2 - 1$.

With 95 percent confidence (assumes a two-tailed test)—if the calculated t_t test value exceeds the table *t* value, then the two averages being compared are different.

n	t value	n	t value	n	t value
6	2.571	26	2.060	45	2.015
7	2.447	27	2.056		
8	2.365	28	2.052	50	2.010
9	2.306	29	2.048	60	2.001
10	2.262	30	2.045		
11	2.228	31	2.042	70	1.995
12	2.201	32	2.040	80	1.990
13	2.179	33	2.037		
14	2.160	34	2.035	90	1.987
15	2.145	35	2.032	100+	1.984
16	2.131	36	2.030		
17	2.120	37	2.028		
18	2.110	38	2.026		
19	2.101	39	2.024		
20	2.093	40	2.023		
21	2.086				
22	2.080				
23	2.074				
24	2.069				
25	2.064				

Simplified *F* Table (95 Percent Confidence)

For comparing sigma from two samples (sizes = n_1 and n_2) (sample sizes equal within 20%).

$$n = \frac{n_1 + n_2}{2}$$

If the calculated F_t value exceeds the table value, assume there is a difference.

n	F	n	F	n	F
6	5.05	26	1.96	60	1.54
7	4.28	27	1.93	70	1.49
8	3.79	28	1.90	80	1.45
9	3.44	29	1.88	100	1.39
10	3.18	30	1.86	120	1.35
11	2.98	31	1.84	150	1.31
12	2.82	32	1.82	200	1.26
13	2.69	33	1.80	300	1.21
14	2.58	34	1.79	400	1.18
15	2.48	35	1.77	500	1.16
16	2.40	36	1.76	750	1.13
17	2.33	37	1.74	1000	1.11
18	2.27	38	1.73	2000	1.08
19	2.22	39	1.72		
20	2.17	40	1.70		
21	2.12	42	1.68		
22	2.08	44	1.66		
23	2.05	46	1.64		
24	2.01	48	1.62		
25	1.98	50	1.61		

Glossary of Terms

Accuracy Accuracy is a measurement concept involving the correctness of the average reading. It is the extent to which the average of the measurements taken agrees with a true value.

Analyze Analyze is the third step in the DMAIC problem-solving method. The measurements and data must be analyzed to ensure that they are consistent with the problem definition and to identify a root cause. A problem solution is then identified. Sometimes, based on the analysis, it is necessary to go back, restate the problem definition, and start the process over.

Attribute An attribute is a qualitative characteristic that can be counted.

Attribute Data Attribute data are data that are not continuous, that fit into categories that can be described in terms of words (attributes). Examples are "good" or "bad," "go" or "no-go," "pass" or "fail," and "yes" or "no."

Averages, Labeling *See* Labeling Averages and Standard Deviations.

Black Belt A Six Sigma black belt has Six Sigma skills sufficient to allow her to act as an instructor, mentor, and expert to green belts. A black belt is also competent in additional Six Sigma tool-specific software programs and statistics.

Chi-Squared Test This test is used on variables (decimal) data to see if there was a statistically significant change in the sigma between the population data and the current sample data. It is done only after the data plots have indicated that there has been no radical change in the shape of the data plots.

Child Distributions This term is used when interfacing with quality department data. A *child distribution* refers to the sample averages and the sigma of multiple sample averages. These are labeled $\bar{x}$ and $s_{\bar{x}}$.

Confidence Tests Between Groups of Data These tests are used to determine whether there is a statistically significant change between samples or between a sample and a population. These tests are normally done at a 95 percent confidence level.

Continuous Data (Variables Data) Continuous data can have any value in a continuum. They are decimal data without "steps."

Control Control is the final step in the DMAIC problem-solving method. A verification of control must be implemented. A robust solution (like a part change) will be easier to keep in control than a qualitative solution.

Control Chart A control chart is a tool for monitoring variance in a process over time. A traditional control chart is a chart with upper and lower control limits on which the values of some statistical measure for a series of samples or subgroups are plotted. A traditional control chart uses both an average chart and a sigma chart. *See* Simplified Control Chart.

Correlation Testing This tool uses historical data to find what variables changed at the same time or position as the problem time or position. These variables are then subjected to further tests or study.

Cumulative In probability problems, this is the sum of the probabilities of getting "the number of successes or fewer," like getting three *or fewer* heads on five flips of a coin. This option is used on "less than" and "more than" problems.

Data Plot Most processes with continuous data have data plot shapes that stay consistent unless a major change to the process has occurred. If the shapes of the data plots *have* changed dramatically, then the quantitative formulas can't be used to compare the processes.

Define This is the overall definition of the problem in the DMAIC problem-solving method. It should be as specific as possible.

DMAIC Problem-Solving Method DMAIC (Define, Measure, Analyze, Improve, Control) is the Six Sigma problem-solving approach that green belts use. This is the road map that is followed for all projects and process improvements, with the Six Sigma tools being applied as needed.

Excel's BINOM.DIST Excel's BINOM.DIST is not technically a Six Sigma tool, but it is the tool recommended in this text for determining the probability that an observed proportional data result is due to purely random causes. This tool is used when we already know the mathematical probability of a population event.

F Test This test is used on variables (decimal) data to see if there was a statistically significant change in the sigma between two samples. This test is done only after the data plots have indicated that there has been no radical change in the shape of the data plots.

Fishbone Diagram This Six Sigma tool uses a representation of a fish skeleton to help trigger identification of all the variables that could be contributing to a problem. The problem is visually shown as the fish "head," and the variables are shown on the "bones." Once all the variables have been identified, the key two or three are highlighted for further study.

Green Belt A Six Sigma green belt is the primary implementer of the Six Sigma methodology. He earns this title by taking classes in Six Sigma, demonstrating a competence on Six Sigma tests, and implementing projects using the Six Sigma tools.

Improve Improve is the fourth step in the DMAIC problem-solving method. Once a solution is identified, it must be implemented. After the solution has been implemented, the results must be verified with independent data.

Labeling Averages and Standard Deviations We label the average of a population $\bar{X}$ and the sample averages $\bar{x}$. Similarly, the standard deviation (sigma) of the population is labeled S and the sample standard deviations (sigma) are labeled s.

Lean Six Sigma A methodology to reduce manufacturing costs through such techniques as reducing lead time, reducing work in process, minimizing wasted motion, optimizing work area design, and streamlining material flow.

Manufacturing Innovation Finding a breakthrough change in a manufacturing process step that takes the process to a higher level or opens up new manufacturing horizons.

Master Black Belt A Six Sigma master black belt generally has management responsibility for the Six Sigma organization. This could include setting up training, measuring its effectiveness, coordinating efforts with the rest of the organization, and managing the Six Sigma people (when Six Sigma is set up as a separate organization).

Measure Accurate and sufficient measurements and data are needed in this second step of the DMAIC problem-solving method.

Minimum Sample Size The number of data points needed to enable statistically valid comparisons or predictions.

n This is the sample size or, in probability problems, the number of independent trials, like the number of coin tosses, the number of parts measured, and so on.

Need-Based Tolerances This Six Sigma tool emphasizes that often tolerances are not established based on the customer's real needs. A tolerance review offers opportunity for both the customer and the supplier to save money.

Normal Distributions A bell-shaped distribution of data that is indicative of the distribution of data from many things in nature. Information on this type of distribution is used to predict populations based on samples of data.

Number *s* (or *x* Successes) This is the total number of "successes" that you are looking for in a probability problem, like getting exactly three heads. This is used in Excel's BINOMDIST.

Parent Populations This term is used when interfacing with quality department data. A *parent population* refers to the individual data and their related statistical descriptions, like average and sigma. These are labeled $\bar{X}$ and S.

Probability Determination This is the likelihood of an event happening by pure chance.

Probability *p* (or Probability *s*) Probability *p* or *s* is the probability of a "success" on *each individual trial*, like the likelihood of a head on one coin flip or a defect on one part. This is always a proportion and is generally shown as a decimal, like 0.0156. *p* is also the probability of an event, such as a defect, in a sample.

Probability *P* In Excel's BINOMDIST, this is the probability of getting a given number of successes from *all the trials*, like the probability of three heads in five coin tosses or 14 defects in a shipment of parts. This is often the answer to the problem. *P* is also the probability of an event, like a defect, in the total population.

Process Flow Diagram The process flow diagram, and specifically the locations where data are collected, may help pinpoint possible areas that are contributing to a problem.

Process Sigma Level This is the formula for calculating process sigma level:

$$\text{Process Sigma Level} = \pm \frac{\text{Process Tolerance}}{2 \times \text{Process Sigma Value}}$$

Proportional Data Proportional data are based on attribute inputs, such as "good" or "bad," "yes" or "no," and other such distinctions. Examples are the proportion of defects in a process, the proportion of "yes" votes for a candidate, and the proportion of students failing a test.

Repeatability Repeatability is the consistency of measurements obtained when one person measures the same parts or items multiple times using the same instrument and techniques.

Reproducibility Reproducibility is the consistency of the average measurements obtained when two or more people measure the same parts or items using the same measuring technique.

RSS Tolerances When establishing tolerances on stacked parts, the traditional method is to use "worst-case" fit, even though the probability of this fit may be extremely low. The RSS method (root sum-of-squares) of establishing tolerances takes this probability into consideration, resulting in generally looser tolerances with no measurable reduction in quality.

Sample Size, Proportional Data This tool calculates the minimum sample size needed to get representative attribute data on a process generating proportional data. Too small a sample may cause erroneous conclusions. Excessive samples are expensive.

Sample Size, Variables Data This tool calculates the minimum sample size needed to get representative data on a process with variables (decimal) data. Too small a sample may cause erroneous conclusions. Excessively large samples are often expensive.

Sigma Sigma, or standard deviation, is a measure of variability used in statistics. It shows how much variation or dispersion exists from the average or mean value. A low sigma value means that the data points are close to the mean. A high sigma value indicates that some data points are further from the mean.

Simplified Control Chart A control chart—traditional or simplified—is a tool for monitoring variance in a process over time. Traditional control charts have two graphs and are not intuitive. Simplified control charts have one graph, are intuitive, and are operator-friendly. *See* Control Chart.

Simplified DOE This Six Sigma tool enables tests on combinations of key process input variables for an existing process to establish optimum settings for the KPIVs.

Simplified FMEA This Six Sigma tool is used to convert qualitative concerns about collateral damage to a prioritized action plan. Unintentional collateral harm may occur to other processes due to a planned process or product change.

Simplified Gauge Verification This Six Sigma tool is used on variables data (decimals) to verify that the gauge is capable of giving the required accuracy of measurements compared with the allowable tolerance.

Simplified QFD This Six Sigma tool is used to convert qualitative customer input into specific prioritized action plans. The customer includes everyone who is affected by the product or process.

Simplified Transfer Function The simplified transfer function shows the contribution of each component to the total variation of an assembly or a process. This allows for a focus on individual components to effect a reduction in total variation.

Six Sigma Methodology The Six Sigma methodology uses a *specific problem-solving approach* and *specialized Six Sigma tools* to improve processes and products. This methodology is data-driven, with a goal of reducing the number of unacceptable products or events. The technical goal of the Six Sigma methodology is to reduce process variation to such a degree that the amount of unacceptable product is no more than 3 defects per million parts. In most companies, the real-world purpose of Six Sigma is to make a product that satisfies the customer and minimizes supplier losses to the point where it is not cost-effective to pursue tighter quality.

Standard Deviations, Labeling *See* Labeling Averages and Standard Deviations.

t **Test** This Six Sigma test is used to see if there was a statistically significant change in the average between population data and the current sample data, or between two samples. This test on variables data is done only after the data plots have indicated that there has been no radical change in the shapes of the data plots and the chi-squared test or *F* test has shown no significant change in sigma.

Tolerance Stack-Up Analysis This is the process of evaluating the effect that the dimensions of all components can have on an assembly. There are various methods used, including worst-case, RSS (root sum-of-squares), modified RSS, and Monte Carlo simulations.

Variables Data Variables data (continuous data) are measurable, generally in decimal form. Theoretically, if you looked at enough decimal places, you would find that no two values are exactly the same.

INDEX

ABOUT THE AUTHOR

Warren Brussee spent 33 years at GE as an engineer, plant manager, and engineering manager. His responsibilities included manufacturing plants in the United States, Hungary, and China.

Brussee is a Six Sigma green belt and has taught Six Sigma classes to engineering and manufacturing teams. These teams excelled both on corporate tests on Six Sigma and in actual implementation of the Six Sigma tools.

Brussee has multiple patents, some of which were the outcome of his Six Sigma work. His teams generated several million dollars worth of annualized savings using the Six Sigma tools.

Statistics for Six Sigma Made Easy! was written to make the statistics involved with Six Sigma user-friendly and to introduce the Six Sigma methodology with a set of simplified tools. The intent of the simplification is to get broader use of this powerful methodology. All the case studies portray real events, with some details changed to protect proprietary processes. The emphasis is on using the Six Sigma tools, not on the theory of Six Sigma. This second edition includes additional examples and a chapter on manufacturing innovation.

Warren Brussee earned his engineering degree at Cleveland State University and attended Kent State toward an EMBA. He has published six technical books, three of which are on Six Sigma, two on the economy, and one on analyzing stocks. He also wrote and illustrated a children's picture book about raising a guide dog puppy for the blind.